British TV and Film in the 1950s:

'Coming to a TV Near You!'

Su Holmes

intellect™
Bristol, UK
Portland, OR, USA

First Published in the UK in 2005 by

Intellect Books, PO Box 862, Bristol BS99 1DE, UK

First Published in the USA in 2005 by

Intellect Books, ISBS, 920 NE 58th Ave. Suite 300, Portland, Oregon 97213-3786, USA

Copyright ©2005 Intellect Ltd

A catalogue record for this book is available from the British Library

ISBN 1-84150-121-2

Copy Editor: Holly Spradling

Cover Design: Gabriel Solomons

Author Cover Image: Spencer Scott

Printed and bound by 4edge Ltd, Hockley. www.4edge.co.uk

FSC Mixed Sources
SA-COC-001695
© 1996 FSC A.C.

Contents

Illustrations

Figure 1: Peter Haigh, presenter of *Picture Parade* (1956, edition 18a)

Figure 2: Derek Bond, presenter of *Picture Parade* (1956, edition 18a)

Figure 3: MacDonald Hobley and his 'quiz' wheel (1956, edition 7)

Figure 4: Opening titles of *Film Fanfare* (1956, edition 15)

Figure 5: Paul Carpenter in *Film Fanfare* (1956, edition 8)

Figure 6: Peter Noble in the 'living-room' set (1956, edition 8)

Figure 7: Peter Noble in his director's chair in the film studio set (1956, edition 19)

Figure 8: John Fitzgerald in the 'viewing theatre' set (1956, edition 15)

Figure 9: John Parsons brings cinema back to the home: 'It has been a truly international week with visitors from Hollywood, Rome and Paris' (1956, edition 19)

Figure 10: Opening titles of *Picture Parade* (1956, edition 18a)

Figure 11: British film director Maurice Elvey in *Picture Parade*'s 'viewing theatre' set (1956, edition 18a)

Figure 12: Diana Decker sings with the orchestra on *Film Fanfare* (1956, edition 7)

Figure 13: MacDonald Hobley 'chats' to British film star Terence Morgan (1956, edition 7)

Figure 14: Peter Haigh interviews Joan Crawford (1956, edition 18a)

Figure 15: Marilyn Monroe faces the media on *Film Fanfare* (1956, edition 23)

Figure 16: Monroe waves to the crowd 'who weren't so lucky as the rest of us' (1956, edition 23)

Figure 17: At the *Yield to the Night* premiere: 'Sabrina was wearing an unusual buckle on the shoulder of her gown' (1956, edition 19)

Figure 18: Gig Young and Jeffrey Hunter introduce the behind-the-scenes story of *The Searchers* (1956)

Figure 19: *Film Fanfare* goes 'behind-the-scenes' on the set of British thriller *Assignment Redhead* (1956, edition 9)

Acknowledgements

I would like to thank colleagues and friends at the Southampton Institute - David Lusted, Karen Randell, Julian Hoxter, Paul Marchbank, Ani Ritchie, Bibi Awan, Nick Rumens and Steve Lannin - for putting up with me talking about 'that 1950s book'. I would also particularly like to thank my friends Sean Redmond and Deborah Jermyn for hearing more about it than most, and for always encouraging me along the way. Equally, thanks go to Pam Cook, Tim Bergfelder and Lucy Mazdon at Southampton University for nurturing my initial interest in the relations between television and film history, and for sharing my enthusiasm for the programmes.

My thanks also to May Yao and colleagues at Intellect for showing an interest in the project, and for seeing it through. To Jenny Hammerton at Pathé Pinewood for allowing me to access the editions of *Film Fanfare*, and to Christine Slattery at the BBC's audio visual archives.

Finally, I would also like to thank my parents, Jenny and Chris Holmes, for trying to remember the 1950s cinema programme, and what it was like when filmgoers first became 'televiewers'.......(They are a bit too 'young', not too 'old').

Copyright:
With thanks to the BBC and British Pathé for permission to use the stills from the programmes.

An earlier version of Chapter Five appeared in *Screen*, 42 (2) summer, 2001, pp.167-187.

An earlier version of the first half of Chapter Seven appeared in *Quarterly Review of Film and Video* 21 (2) April-June, 2004, pp.131-147.

An earlier version of the second half of Chapter Seven appeared in *Historical Journal of Film, Radio and Television* October, 2001, pp. 379-397.

Introduction

Cinema in the Home: 'The Entertainment of the Future?'

The year is 1952. Britain's most popular film fan magazine, *Picturegoer*, has just published its survey on 'Films and TV', which enquired: 'What effect - if any - is television having upon the film habits and appetites of British picturegoers?'[1]. But at the same time (and on the same page), the magazine also reports news of a TV cinema programme and explains: 'As the first regular link up between the BBC and your local cinema... it may shape the entertainment of the future'. Now cut to 2002, and the popular BBC1 series *Alistair McGowan's Big Impression* (2001-), featuring impressions of contemporary media celebrities. One sketch involves an impersonation of Jonathan Ross, presenter (at the time of writing) of the BBC's Film 2004 programme. A key feature of the sketch involves 'Ross' complaining about the lack of priority given to his programme, demanding in particular that the BBC 'Put me on earlier!' (The programme usually goes out at 11:35 pm). He emphasises how people arguably seem more films per year than they take holidays, yet holiday programmes take up more time in the television schedules than cinema reviews.

The differences between these constructions of the cinema programme could be explained by reflecting on the decline of the cinema as a mass medium. *Picturegoer* could not necessarily foresee in 1952 that the impending shifts in cinemagoing would make it unlikely that the cinema programme would ever really represent 'the entertainment of the future', as the comedy sketch above confirms. But *Picturegoer*'s comment points to more than simply the dangers of forcing a retrospective point of view onto an earlier period. Crucial here is that its news was characterised by excitement about this 'modern' media 'synergy', not simply because it was hiding behind a 'naïve' inability to take a more 'realistic' look at the future of the cinema as a mass medium, but because in 1952 the cinema programme was representative of a different set of cultural, economic and technological possibilities for the relations between cinema and television. In fact, it is not an exaggeration to suggest that this was a time when these programmes were at the centre of the emerging relations between British film culture and television. In 1956 the BBC's *Picture Parade* appeared on the screen by announcing itself as 'a weekly montage of news from the world of the cinema' and it is by exploring this 'world' - television's world of cinema - that there exists a further lens through which we can consider the historical interaction between the media. In a period when television is perceived to have been cinema's biggest rival and at the very least, a prime factor in hastening its decline, how did it simultaneously play a role in promoting the medium, so that its films and stars, in fact its entire culture, pervaded the domestic sphere? What are the implications of these programmes for our understanding of the historical interaction between cinema and television and their cultural identities at this time? How did television offer perspectives on the cinema which were shaped by the aesthetic, technological and cultural specificities of its form, and how did these contribute to the particular experience of 1950s film and television culture?

This book explores this cultural role by tracing the development of the cinema programme on British television in the first decade of its existence: 1952-62. The primary focus here is on the three key series of the time, *Current Release* (BBC, 1952-3), *Picture Parade* (BBC, 1956-62) and *Film Fanfare* (ABC, 1956-7), and the book examines their emergence, development and change within this early period of interaction between cinema and television. The fact that these programmes are so central in understanding their historical relations here points to what I argue is the genre's 'special' or unique status in the 1950s. One of the main reasons for this is that the roles of cinema and television (as media industries but crucially as sites of screen entertainment), effectively came together at this time in such a way that is unique to the decade. Television's growing status as a mass medium developed throughout these years, and although this certainly hastened the cinema's 'decline', the cinema still remained a central part of cultural consciousness and experience for much of the 1950s. This created a context in which the media temporarily shared a status as forms of 'mass' screen entertainment, something that is often overlooked in conventional perspectives of their relations at this time. Particularly after the advent of a British television's second channel in 1955 (ITV), it is not surprising when viewed from this perspective that television coverage of the cinema became a key site in the competition for audiences. It was part of a shared, and still 'everyday' culture that defined the existence of the cinema in such a way that is different from today. While the title of the collection *Hollywood in the Age of Television* (Balio, 1990) is intended to signify the cinema's continued existence in a changed media 'age', there is a sense in which my focus here conceives of these programmes as television in 'the age' of the cinema.

Crossing the 'Divide': Constructing Film and Television History

The cinema programme was a key prime-time genre on television in the 1950s in such a way that has not been repeated since. Central here was also television's attitude toward the cinema: its utter deference and excitement, the unflagging celebration of its glamour, importance and charm. Television's attempts to borrow and capitalise on the cinema's 'glamour' are central here. Despite the popular and pervasive picture of decline, it is perhaps the case that the cinema was nowhere more 'alive' in the 1950s than on television - the small screen space where it seemed to revel in a climactic celebration of its last gasp as a mass medium.

In straddling both media, the early cinema programme has a multitude of implications for connecting with many areas of film and television history. It offers, for example, a further (and in some ways quite literal) 'window' on 1950s British film culture in such a way that is very different from that offered by the films themselves, or from the written narratives found in established histories. The window of the cinema programme depicts a week in, week out rhythm of film culture - its stars, films, studios and debates - which presents an intriguing angle on the contours, look, energy and 'feel' of British film culture in the 1950s. To a certain extent, it undoubtedly has this in common with the fan magazine which, aside from radio, was clearly a key precursor in the intertextual construction of the cinema. Evidently, however, television's 'window' on the cinema is made possible precisely because of their shared status as forms of screen entertainment (which of course also structures the culturally charged nature of their media relationship). It is the dialogic thrust of this history, telling us as much about the cinema as it does about television, which characterises the critical focus and methodological approach of this book.

While the cinema programme's unique status in this period emerges from the ways in which the cultural centrality of cinema and television momentarily coalesced, it is also the case that its relations with the cinema had a contradictory status, effectively looking two ways at once. Although undoubtedly working to keep the cinema's 'mass' status alive, it could only do this by delivering it to the increasingly mass audience for television. The programmes, then, were equally representative of the domestication of film culture, and are also positioned as transitional - 'caught' between, and reflecting on, the irrevocable shifts in the consumption of screen entertainment. In her discussion of the contemporary domestic consumption of film, Barbara Klinger (1998a) has described how the 'specifically "cinematic" subject has ceased to exist'. Yet despite film studies' continued tendency to analyse the medium 'almost exclusively as a phenomenon of the big screen' (Klinger, 1998b, p.4), this concept of the 'spectator/viewer' is clearly not a recent shift (given that films have been consumed in the home since the 1950s). However, we seem to have comparatively little sense of a historical foundation on which to consider this domestication of cinema, and the dynamics of its circulation, exhibition and reception in the domestic sphere. Although different to the consumption of full-length feature films in the home (partly because one of its aims is to encourage us to venture out to the public space of the cinema), the early cinema programme is a historical site for considering elements of this process, and the technological, aesthetic and cultural parameters in which it developed.

The subject and approach of the book clearly involves considering cinema and television together, and this is something which poses particular critical and methodological challenges (as well as pleasures). What can still be described as a disciplinary 'divide' between film and television studies has had considerable implications for the writing of film and television history. For example, Charles Barr commented in 1986 that 'Television's past even now remains relatively unchronicled, and is not often correlated with film history' (p.222). Although there are now exceptions to this (see Hill and McLoone, 1996, Stokes, 1999), this situation has changed little where the British situation is concerned. This is quite different to the American context. Shaped by the historical turn in both film and television studies, the late 1980s onwards saw substantial works on the relations between Hollywood and broadcasting (Hilmes, 1990, Balio, 1990, Boddy, 1990, Anderson, 1994). In 1989, for example, Michele Hilmes pinpointed the direction of this work by arguing for 'a new perspective on the film/broadcasting relationship... one which goes back to primary materials to recast a different historical narrative in different terms' (1989, p.39). Existing work had insisted on Hollywood's hostility toward radio and television, suggesting that it was reluctantly forced to co-operate only after a protracted period of competition. Yet in revisiting archival material, the revisionist studies discovered how Hollywood and broadcasting had been interlocked at both economic and textual levels for decades (Anderson, 1991). From this perspective, the emergence of television only served to reinforce a complex media alliance that was firmly established with radio in the late 1920s.

My suggestion that cinema and television need not primarily be cast as 'enemies' in the 1950s is clearly not radical from the perspective of these later arguments. But it is more suggestive in relation to the British context. Certainly, this is in part due to the different industrial and institutional infrastructures of cinema and television in Britain, as well as the greater difficulty of locating and accessing the archival sources through which this history might be told. Yet it is also shaped by the ways in which this media history has been constructed. The work which does exist has tended to fall into two areas: firstly, a discursive approach which considers how British films represented television (Barr, 1986, Stokes, 1999), and secondly, discussion of the disputes surrounding the sale of feature films to television (Buscombe, 1991). Referenced in overviews of British cinema of the period, as well as histories of British broadcasting, the presence of the second narrative have been more pervasive, although an article by Ed Buscombe (1991) represents the only sustained study of the issue. Entitled 'All Bark and no Bite: the Film Industry's Response to Television', Buscombe's article focuses on FIDO, the Film Industry Defence Organisation, which was set up in 1958 to purchase the rights to feature films to prevent their sale to television. That FIDO was indicative of a very antagonistic attitude toward television is not in doubt here. The problem lies in the construction (and canonisation) of this history, and the way in which it has come to represent the early relations between British cinema and television. Buscombe's research leaves us with a sense of the British film industry's hostility toward, and distrust of the new medium, explaining how, when faced with the competition from television, its instincts 'were always restrictive rather than expansionist' (Buscombe, 1991, p.206). When television is mentioned in more general surveys of British cinema in the 1950s, the historical contours of the situation are quickly sketched for the reader by references to FIDO, the disputes over feature films, and the now familiar conceptions of hostility (Perry, 1977, Hill, 1986, Park, 1990, Murphy, 1992).

This perspective has acquired consensus in the context of the critical neglect of (and disdain for) British cinema at this time, although more recent work has certainly sought to address this (see Geraghty, 2000a, Stafford, 2001). Geoffrey Macnab explains how 'The decade is held in low esteem by film historians who are wont to lump every film together and dismiss the whole sticky mess' (1993, p.219), while James Chapman argues that as the '"doldrums era" of British cinema', the 1950s are 'usually regarded as a period of stagnation and subsequent decline, as film production became increasingly standardised and stereotyped, and cinema attendances began to fall off' (1998, p.66). This emphasises how the critical disdain for British cinema in this period is implicitly enmeshed within the wider sense that it signifies a general slide toward 'decline' - whether in terms of the fortunes of the British film industry, the aesthetic and artistic energy of its production, or the status of the cinema as a mass medium. From this perspective, we are left with the impression that television largely functioned as part of the 'mess' Macnab describes. In fact, Barr boldly associates the decline of a 'coherent British domestic cinema' with the advent of British television (1986, p.207). This clearly taps into much wider discursive constructions of British cinema beyond the 1950s and the degree to which it is defined in parochial, insular and 'protective' terms, particularly, of course, in comparison with Hollywood (Cook, 1996, p.1). It is not insignificant here that this in turn works to further fuel the perception that the British film industry's response to television was 'predictably' 'restrictive rather than expansionist' (Buscombe, 1991, p.206).

The prevailing assumptions about the early antagonisms between British cinema and television are based not only on particular conceptions of the British film industry, but also British television. Although there are signs of change in so far as archival research is increasingly being undertaken on commercial channel ITV (see Thumim, 2002a, Wegg-Prosser, 2002), the BBC has undoubtedly dominated the writing of British television history. This is clearly in part because of its institutional role in developing the medium and forming an economy of public service, but it is also again shaped by the availability (and accessibility) of archival sources. Particularly in histories and popular perceptions of the 1950s, the BBC is widely discussed as elitist, paternalistic and didactic, a perception which fuels the image of antagonistic and 'uncomfortable' relations between cinema and television. It is suggested, for example, that it was not simply the film industry's refusal to sell their feature films to television which is significant here, but equally the BBC's elitist rejection of the commercial cinema - their fears that it might conflict with the institutional responsibilities of public service. Briggs and Buscombe, for example, both quote from the Head of the Television Service, Norman Collins, and his assurance to the BBC's board of Governors that: 'It is no part of the

Corporation's intentions to convert the BBC Television Service into a home cinema, showing mainly commercial films. It has a far more serious responsibility' (Buscombe, 1991, p.22, Briggs, 1975, p.275). This perception permeates more general accounts of BBC television (Wyndham-Goldie, 1977, Crisell, 1997), but it is difficult to reconcile this attitude with the BBC's engagement with British film culture, not only from the advent of television but, as with the relations between Hollywood cinema broadcasting, right back to radio in the late 1920s when the Corporation's coverage of film culture began. If the cinema were simply regarded by the BBC as a 'low-brow' mass leisure activity, then surely regular radio, and then later television, programmes promoting its wares would be most undesirable. It is true that the BBC's relations with film culture were a site of constant debate, conflict and negotiation where the ethos of public service was concerned. Yet this was also an arena in which the BBC saw many possibilities to connect with the audience, to educate and inform but crucially, to celebrate film, to entertain. Broadcast coverage of film clearly represented a parallel narrative to the history of British film culture, whether at the level of institutional relations, texts or reception.

The different images we now have of the British and American contexts at this time are shaped as much by the construction and focus of existing accounts, as they are by the historical relations which occurred. In this respect, as the work of scholars such as Hayden White (1975) has suggested, history is 'fundamentally an interpretative project' (Spigel, 2001, p.12). As Lynn Spigel explains:

> History is a kind of knowledge based not only on the historian's subjective determinations regarding evidence but also on the conventions of writing that govern other kinds of textual production... It is the process of interpretation - and the ways in which we use evidence to produce an argument - that is at stake

(Ibid).

My own consideration of the cinema programme in the 1950s is clearly also situated within, and shaped by, the 'interpretative project' of history. This includes the use of available historical sources discussed below, but more generally, the desire to challenge the negative accounts of the relations between British film culture and television at this time. Studying the early relations between Hollywood and broadcasting as a student (I recall approaching the exam question 'Was television ever really a threat to Hollywood?'), I had found the debates, and particularly the 'rediscovery' of archival sources, fascinating: going back to a time when broadcasting and cinema were 'unfamiliar' with one another, and tracing their gradual encounter anew. I remember thinking that it was a pity that there wasn't a similar narrative to construct of the British context - as the silence around the area seemed to suggest. Certainly, as detailed in Chapter One, the British situation *was* different, not least of all due to the different institutional and industrial infrastructures of film and television, but this is not the same as suggesting that there was (is) no narrative here to tell at all.

While this revisionist narrative offers an important undercurrent in the contribution the book aims to make to the histories of British television and cinema, it explores one of the most popular genres of early British television and concurrently, its role in the construction of 1950s British film culture - its filmmaking, fortunes, technologies, genres and stars. In terms of television, the book is interested in the ways in which the programmes can reflect upon the broader development of the medium at this time in terms of institution, aesthetics, technology and viewing cultures. In this respect, it combines what Jason Jacobs describes as the 'macro-overview of broadcasting history' with the more local analysis of a specific genre or set of texts (2000, p.9). This has become increasingly central to historical approaches to television (see Corner, 1991, Thumim, 2002a), but it remains the case that it is still relatively rare to give a particular early genre the kind of detailed analysis offered here (particularly a form of non-fiction or magazine programming). Yet the book is deliberately more expansive than this given that its history is also situated within, and consistently connected to, the parallel realm of the cinema, and the spheres of film culture and the film industry play a larger role than simply that of a monochrome image on the television screen. This is particularly so given that even the historical work on Hollywood and broadcasting has tended to privilege institutional and industrial perspectives (Balio, 1990, Boddy, 1990, Hilmes, 1990, Anderson, 1994), and the implications of television's coverage of the cinema sometimes seems lost within a welter of information on industrial and economic relations. These structures are clearly crucial, but they don't on their own suggest a sense of what these developing media relations looked like - how they participated in the domestication of film culture, how they were played out for popular cultural consumption, and how they formed textual spaces which audiences consumed at this time of change in the history of screen entertainment.

Nevertheless, Hollywood film culture is in itself important here. Not only was there the strong presence of American film companies in Britain which represented keen contributors to the cinema programme, but the remit of these series - covering films being exhibited at the cinema - necessarily means ranging across national borders. Unsurprisingly, given its economic and cultural power, Hollywood cinema was fundamental to the texture of these programmes, and the case studies discussed here often focus on Hollywood films, companies and stars. Necessarily moving beyond a comparative analysis of the British and American contexts, this focus also enables a further perspective on the circulation of Hollywood film culture in Britain at this time (see Swann, 1987).

It is in this respect necessary to qualify the use of the term 'British film culture' here. In relation to the constructions of British cinema indicated above, a 'British film culture' has often (problematically) been conceived in anachronistic, protective, insular and 'patriotic' terms (Cook, 1996, Street, 2001), although the concept of what constitutes a 'British film culture' is complex and contradictory, and is always be subject to cultural change, struggle and debate (Street, 2001, p. 6). Given that the programmes here are a moving texture of British, Hollywood and (as discussed in Chapter Seven), 'Continental' films, my use of the term is not intended to refer to the national specificity of a cinema (although this in itself may reflect back on the fact that conceptualising a 'distinctive British film culture' is particularly complex in the face of the market domination of Hollywood) (Street, 2001: 10). Nor is my use of the term intended to signify a form of 'ideological framework' which, in any given period, structures the production and consumption of films (Ryall, 1986, p.2). For reasons of simplicity, in the context of this study and the focus on the cinema programme, I use the term 'British film culture' to signify the films (and stars) circulated, exhibited and consumed in Britain in the 1950s. In this sense, the term 'culture' here could be taken out of its more immediate (media) context to reference Raymond Williams' famous definition of 'culture' as constituting the 'ordinary' and the 'everyday' (Williams, 1983, p.87). Apparent in this period was an endless flow of 'new' releases, established/ 'up-an-coming' stars and studio news which, despite the changing status and situation of the cinema, was in many ways much like any other. These were the films which represented the popular face of filmgoing, which were advertised on billboards and buses, and which featured in fan magazines, and in radio and television programming. As Street argues, the intertextual, construction of the cinema has always represented a shaping practice in how our sense of 'film culture' is constituted, and how the cinema is actively 'understood' (Ibid). However, in the 1950s, television - as part of this media context - was still 'new'. This everyday 'ebb and flow' (Street, 2001: 10) of the cinema's existence in Britain, is at the same time rendered 'unfamiliar' by its representation through the lens of television - both for the audience at the time, and in terms of the perspective of the researcher. It is this duality or contradiction that structures much of intrigue or fascination with these programmes.

In discussing the development of television programming, John Corner has suggested that a 'primary factor in the formation of generic styles was the search for the distinctively "televisual", which perhaps reworked from cinematic, theatrical...[or] radio... precedents, but which used the medium to its best possible advantage' (1991, p.13). Clearly, in studying the development of television genres there is a need to balance perceptions of their innovation and specificity with an understanding of their heritage and precursors in other media forms. Media coverage of film was clearly not without precedent, and in analysing television's intervention here, every attempt has been made to contextualise the perspectives offered by other media forms (particularly the fan magazine and radio). Within this context we can consider how television adapted or elaborated on conventions established in other media but, crucially, how it also offered new perspectives on the coverage of cinema. In short, how it made film 'televisual', and what this might have meant in the 1950s.

As we shall see, from its earliest days the cinema programme was criticised for being simply a 'bargaining' tool with which television hoped to secure feature films in return. However, given the wider fetishisation of the broadcast feature film in the work on the industrial and aesthetic relations between the media (Buscombe, 1991, Belton, 1992, Maltby, 1983, Lafferty, 1990), it is worth noting here that broadcasting emerged only twenty or so years after the cinema itself. It has long since been in its coverage of the cinema that broadcasting has exerted a shaping influence on our access to film culture, and our understanding of what this actually is. In the extensive range of work which has increasingly given attention to the wider construction of the cinema - in promotion, exhibition, fan cultures and intertexts - it seems hardly necessary to note now that there is 'more to cinemagoing than seeing films' (Morley, 1989, p.26). My focus here is not a textual analysis of the films from the period, but rather an exploration of the audio-visual material which functioned to construct their circulation. Although issues of reception are only part of this context, it is worth emphasising that although we often can't meet with audiences from the past, we do have access to the material which they read and viewed as part of everyday life (Jenkins, 2000, p.169).

In the Archives: (Re)constructing the 'Ghost' Text

This interdisciplinary focus shapes my theoretical and methodological approach, and in particular, the historical sources that are used to reconstruct the culture of the 1950s cinema programme. While cinema historians have a varied, although incomplete history of film from its earliest days, a substantial part of television's past inhabits what John Caughie describes as a 'dim pre-history' in which programmes do not exist in recorded form (2000, p.9). Institutionally, technologically and culturally, early television programming was regarded as live and ephemeral, and there is almost a complete absence of audiovisual material from the period before 1955 (Jacobs, 2000, p.4). The BBC programmes have been reconstructed from evidence at the BBC Written Archive Centre (WAC), chiefly internal memos, documents, press cuttings and the programme scripts themselves. The BBC's *Current Release* and *Picture Parade* were broadcast live. Although certain editions of both programmes were telefilmed, there are no surviving editions of *Current Release* (the programme analysis of which is based entirely on scripts). There is limited audio-visual access to the BBC's *Picture Parade*, offering a combination of programme footage and scripts. The paucity of audio-visual material here is a limitation when it comes to understanding visual style, aesthetics and address, although it is perhaps more appropriate (and less negative) to suggest that research context demands a reformulation of more traditional notions of textual analysis (Jacobs, 2000, p.14).

In discussing programmes 'that do not exist in their original audio-visual form but exist instead as shadows, dispersed and refracted among buried files, bad memories, a flotsom of fragments' (Ibid), Jacobs describes a similar approach in his excavation of early British television drama. Capturing the simultaneous 'presence' yet absence of these programmes, Jacobs refers to these as 'ghost texts'. The institutional sources offer a very strong and tangible sense of the day-to-day running of the programmes, as well as the factors and debates which influenced their development, production and consumption. Although this evidence is particularly valuable in offering an 'insider's' view of the decisions, problems and planning which surrounded the cinema programme from its earliest days, there is nevertheless often the impossibility of reaching out to 'touch' them, or fixing the text in one's gaze. It is also worth acknowledging here that such evidence is not without its limits. Not only is it shaped by the conventions and form of 'institutional' correspondence (and presumably not everything was recorded on paper or filed), but it also tends to privilege the perspectives of those in positions of authority (such as the Programme Controller) who may be more removed from the actual production of the programmes themselves (Jacobs, 2000). The situation with archival sources is almost the reverse where the commercial channel ITV is concerned. Partly reflecting the extent to which liveness was declining as the primary aesthetic and technological conception of television, the majority of ABC's *Film Fanfare* were recorded on film, and over thirty editions survive in the British Pathé archive at Pinewood Studios. While offering a much wider access to visual textures of the series itself, there is simultaneously little contextualising archival material to accompany ITV's involvement with the programme - potentially positioning the viewer as less 'insider' than as a viewer (with all the retrospective difficulties this involves) (see Thumim, 2002a, p.2). It is partly as a result of this that the BBC dominates this account of the institutional history and development of the genre, although this can partly be justified by their more prominent role in this process at this time. However, in spanning the end of the BBC's monopoly and the introduction of commercial television, the development of the medium's relations with film culture span an important time of transition in the institutional and cultural status of British television. As such, the production and circulation of the genre simultaneously offers insights into wider issues in television's development in the 1950s on a number of different levels.

The interdisciplinary nature of this study also demands the use of a wider range of archival sources which enable a more detailed consideration of the film industry's involvement in the cinema programme. Although representatives of the film industry appear as key 'players' in the narratives of the BBC's institutional sources (such as minutes from meetings), it is important that the film industry's trade

press - such as *Kine Weekly*, *The Daily Film Renter* and *Today's Cinema* -
included daily discussion of the programmes throughout their run. This offers a
revealing counterbalance to the institutional (and broadcast) perspective of the
BBC evidence, yet it is material similarly structured and mediated by its own
contextual form. It is worth remembering that the trade press is written not simply
by, but crucially also for, the film industry. Preoccupied with the declining cinema
audience, television was unsurprisingly at the forefront of discussions in the trade
press at this time and its construction shifted between ally and enemy as the film
industry faced an increasingly uncertain future. Within this context, there was
clearly an attempt to exercise a discursive control over the medium - that is, playing
down its threat to the cinema, while playing up notions of hostility and 'battle'. In
terms of the cinema programme, this often lead to a fascinating disparity between
the BBC evidence, the programmes themselves, and the day-to-day narration of the
interaction in the trade press. In contextualising the wider circulation of the
cinema programme, other important sources include the general press, and
Picturegoer fan magazine, each of which are used to trace elements of popular
reception.

In terms of periodisation, I refer to the series here as all examples of the 'early'
cinema programme, which may be taken to imply their existence as part of early
television. There is some contestation over what constitutes the 'early' phase of the
medium, particularly given that it first began transmissions in Britain in 1936
(before closing down for the war in 1939 and re-opening in 1946). Jacobs' study of
television drama defines early television as occupying the period up until 1955 - a
chronology based on the distinctive features of a single channel, particular notions
of public service, and the predominance of live transmission (2000, p.5). Hence, if
following Jacobs' definition, the book spans the early development of television, as
well as its transition into a period of greater maturity and expansion. Yet - with the
cinema programme only beginning in 1952 - it also seems reasonable to point out
that what is 'early' necessarily varies between different genres, although it is clear
that this doesn't fundamentally change the context of television's development as a
mass medium. However, in terms of periodisation where the cinema programme is
concerned, there is also the need to account for the parallel sphere of the cinema.
The suggestion that there is an 'early' cinema programme as distinct from a later
form is linked to the specificity of the cinema's cultural role in the 1950s which is
attached to its status as a 'mass' medium (with the qualifications of 'decline'
already discussed). As a result, approaching periodisation solely within the
economic, institutional, technological or cultural parameters of television is not
entirely sufficient here. The time span of the study (1952-62) represents what I
have established to be the rise and fall of this popular genre within the dual
contexts of cinema and television. This takes the reader from the emergence of
Current Release to the end of *Picture Parade*, by which time the shifting cultural
roles of cinema and television meant that genre was subject to significant change.

'Coming Soon' across the decade: The Structure of the Book

The book is divided into eight chapters, the progression of which is intended to reflect the development of the cinema programme in this period, and its relations with the changing contexts of both television and cinema. Chapter One sets the wider context for the analysis of the programmes by considering the ways in which the historical relations between film culture and television in Britain differed from those in America, and the institutional and economic reasons for this. It indicates aspects of the wider interaction between British cinema and television in the 1940s and 1950s which include the debates surround Cinema-TV, the disputes over the sale of feature films to television, and the involvement of film companies with the advent of ITV. This survey also includes consideration of the wider impact television had on cinemagoing in this period, and the implications of the cinema programme within this cultural context.

In exploring the institutional origins of the genre and the factors behind its emergence, Chapter Two establishes the argument that the early cinema programme was not simply a bargaining tool or 'favour' to the film industry with which the BBC hoped to obtain feature films in return. Certainly representing the conventional response to the genre in the press at the time, these discourses played a key role in constructing its (low) cultural value which, in view of its promotional connotations, it arguably still has today. In contextualising the emergence of the cinema programme, it is necessary to consider elements of the previous relations between radio and film, as well as the tentative origins of the televised cinema programme in the late 1930s (although no programme actually emerged at this time, it is in this period that the BBC sought to develop the idea). The planning and development of *Current Release* in the early 1950s then demonstrates how the key issue dominating the negotiations between the BBC and the film industry were the extent to which they had differing investments in the idea of the cinema programme, and hence different perceptions of what it should 'be'. The BBC required the series to be a 'worthwhile' and respected element of their programme repertoire while, approached as promotion and publicity, the film industry required to series to function as an effective marketing tool. Within the context of these issues, Chapter Three moves on to offer a textual analysis of *Current Release* and secondly, a small-scale study of its reception. This involves examining, on a limited scale, how the audience may have used *Current Release* in the context of changing attitudes toward cinemagoing and an economy of early TV viewing - traces of reception which also reflect back upon the programme's 'difficult' status at the intersection of promotion and public service.

The next three chapters are concerned with the later *Picture Parade* and *Film Fanfare*. Chapter Four explores how it was during the period 1956-8 that the cinema programme was apparently at the peak of its institutional and cultural significance. It is at the start of this period that, with the cinema programme now the primary locus of interaction between the media, the relations between the film industry and television become institutionalised. This was partly shaped by the advent of ITV and its expansion of the opportunities for the coverage of film culture, the competition for the TV audience, and improved relations with the film industry.

Chapters Five, Six and Seven move away from a focus on the wider development of the cinema programme to explore particular areas and case studies in its mediation of film culture. Chapter Five examines the construction of what was the very nucleus, heart or fabric of the genre - not so much the films, as their stars. Here, I consider the historical emergence of the film star appearance on television, and the discourses which surrounded the shift of film stars to the small screen. These discourses are then rendered more tangible through an analysis of two key case studies focusing on Hollywood stars and their visits to the country in 1956: Joan Crawford (on *Picture Parade*), and Marilyn Monroe (on *Film Fanfare*). The aim here is to examine how these appearances were shaped by the aesthetic, technological and cultural form of television, and its developing modes of domestic address. At the same time, these appearances can be considered as interacting with the intertextual construction of their particular star images at this time. Chapter Six examines two further key areas in television's construction of film culture: First, the coverage of the film premiere, and second, the 'behind-the-scenes' perspective on filmmaking. The discussion of the film premiere - while still considering the representation of the star - is particularly interested in the issues surrounding the technological construction of these events, and focuses predominantly on the concept of the live outside broadcast. Consideration of the 'behind-the-scenes' perspective in filmmaking explores how this approach was developed by television, and in particular, it returns to the industrial influence of Hollywood cinema in shaping the relations between film and television.

Chapter Seven returns to a key and established narrative in the construction of film and television history in the 1950s: that the cinema aimed to differentiate itself from (and compete with) television through a partial 'reinvention' of its aesthetic, technological and cultural identity. This has chiefly been conceived in two ways. The cinema's use of innovations in film technology (colour, widescreen) and innovations in film content (images and narratives which pushed the boundaries of film censorship). However, again a product of treating film and television histories as autonomous (and playing up conflict between individualised media), the cinema programme complicates this argument by indicating how television played an active role in promoting these shifts - whether through its introduction of Cinerama, routine previewing of CinemaScope films (which was of course not without its technological 'problems'), or discussion of 'X' -rated and 'Continental' features.

Chapter Eight addresses the period 1959-62, the boundaries of which are marked by the beginning and the end of the 'new' *Picture Parade*. The programme returned after a break in 1959, but viewers in the late 1950s must have found it much changed. Gone was the overwhelmingly populist, celebratory and deeply affectionate immersion in British film culture which had characterised the early cinema programme. It is because of this that the end of the 1950s and early 1960s can be seen to represent the 'decline' of the cinema programme as a mainstream television genre. Given that it still appears on television today, it may sound perplexing to describe the genre as declining some forty years ago, yet it is at this point that the early cinema programme indeed becomes a template for its successors. As epitomised by *Picture Parade* and *Film Fanfare*, the popular cinema programme was outmoded in its celebration of the cinema as a mass medium, part of the fabric of everyday life. As this suggests, a key factor shaping this shift was the changing cultural role of the cinema, although it is equally important to consider other factors such as the institutional development of television, and broader cultural and social changes as a new decade dawned. As the 1950s drew to a close, these developments in the cinema programme foreground the historical specificity of its identity in its earliest years. The journey through this period is the objective of the book.

Notes
1.   Picturegoer, 'Films and TV Survey', 9 February, 1952, p.5

Chapter One

Broadcasting It:
Approaching the Historical Relations between Cinema and Television

This chapter has two related aims. First, given that the cinema programme did not exist in isolation from other spheres of interaction between British cinema and television, it contextualises the wider activity and negotiation occurring at the time. Second, it examines *how* and *why* the early relations between British broadcasting and film culture have differed from those in the US and, more specifically, the ways in which this created a particular space for the cinema programme to emerge as the primary site of interaction between British cinema and television in the 1950s.

It has been suggested in the preceding pages that if there are considerable contrasts in historical accounts of the interaction between cinema and broadcasting in Britain and America, then this has as much to do with the *construction* of these histories as it does with the actual historical relations which occurred. In short, it is fair to suggest that this disparity is a combination of our understanding of this history, as well as the different national contexts in which cinema and broadcasting have developed. Hilmes' observation that, by the late 1930s, 'radio virtually *became* Hollywood' (1989, p.41), or Anderson's argument that the 'history of American TV is the history of Hollywood TV' (1994, p.12), immediately emphasise the different institutional and economic structures in place here. A similar conception could not be applied to history in the British context, however it might be constructed or explained. Only with the more recent establishment of British television as a key economic force in British film production, not least of all with the advent of Channel Four's initiative *Film on Four* which produced films with both a theatrical and television release (Hill, 1996, Stokes, 1999, p.44), have critics felt compelled to make similar comments about a truly 'symbiotic' relationship between the media. As John Caughie and Kevin Rockett argue, for example, 'since at least the early 1980s it has been impossible to discuss British cinema without also discussing British television' (1996, p.6). In this respect, British television's involvement in film production has emerged less out of purely commercial imperatives than an ethos of public service (Hill and McLoone, 1996, p.2). For some, this has fostered an economic and aesthetic framework in which the films are seen as 'lacking in cinematic values' - effectively falling short of being 'real cinema' (Hill, 1996, p.166). These views clearly arise from wider prejudices surrounding *technological and aesthetic* comparisons of the cinema and television. However, they also reflect back on perceptions of the interaction between British television and cinema as being in some way more 'parochial', less successful (and in fact less interesting) than the history of the cinema's relations with broadcasting in the American context. Particularly when it comes to historical analysis, this perception ultimately pivots on the basis of more direct forms of economic or institutional interaction - in contrast to forms of *textual* interplay which may, or may not, be directly linked back to economics. Martin McLoone has argued that:

> [T]here are really two relationships at stake. First, there is the relationship between cinema and television as institutions and second, between the respective media of electronic imaging... and film. These are not, to my mind, the same thing, though they are often talked about as if they were inter-changeable oppositions...[original emphasis]

(1996, p.83).

While the period under consideration here points to similarities between the British and American situations, it is precisely the different institutional infrastructures of television and cinema in Britain which mean that forms of *textual* partnership are more important. As discussed below, in both Britain and America, cinema interests were initially prevented from developing controlling interests in either broadcasting institutions or technologies. The key difference with the American context was that the Hollywood film industry were able to become much more involved in the production of broadcast programming from an early stage. When considering the British context, it is for this reason that it becomes important to make a distinction between economic interaction and what we might call *textual collaboration*, as it was this, in the form of the cinema programme, which played the crucial role. Nevertheless, despite certain similarities, it is also difficult to make direct comparisons between Britain and American given the different structures and scope of their respective film and television industries (see Stokes, 1999). Of course, this study could equally begin by invoking a comparison with other European contexts (were similar early research on the relations between cinema and television available here). However, the narrative of Hollywood and broadcasting, as with Hollywood cinema itself, is undoubtedly the most 'known' history in this field, and it is implicitly this to which the Britain situation is compared. Furthermore, the historical centrality of Hollywood film culture in Britain shaped the British cinema programme, a context which makes the Hollywood film companies part of its historical narrative.

'Controlling' Television?: The Film Industry and Television in the US and UK

In both Britain and the US, the organisation of television was shaped by the existing structures of radio, thus developing under an ethos of public service in Britain, and the commercial networks in America (Stokes, 1999, p.24). Later to develop under the infamous moral principles of the first Director General of the BBC, Lord Reith, the government deemed that the wavebands necessary to broadcast radio be regarded as a 'valuable form of public property' (Scannell, 1990, p.11). The scarcity of wavelengths, as well as the perceived social, cultural and political power of radio, all shaped the belief that 'the operation of so important a national service ought not to be allowed to become an unrestricted commercial monopoly' (Scannell, 1990, p.13), and the government ultimately awarded a monopoly, in the context of public service, to the BBC. In differing from the highly complex, competitive and commercial network of radio's infrastructure in the US (see Hilmes, 1990), it was precisely this concept of a public service monopoly which made the key difference in terms of the film industry's interaction with broadcasting, and the types of partnership which prevailed.

In contrast to the conventional narrative of Hollywood's approach to broadcasting as moving through 'complacency, competition, co-operation' (Anderson, 1991, p.85), revisionist work has emphasised how Hollywood's failure to establish ownership or control of radio and television was shaped by more pragmatic reasons. The major Hollywood film companies consistently sought to get in on the 'ground floor' of both radio and television - in terms of developing alternative broadcast technologies or operating network stations (White, 1990, Hilmes, 1990). A combination of factors prevented this, not least of all opposition from broadcast interests and the regulations concerning monopoly ownership - as enforced by the Federal Communications Commission (FCC). To place radio in the hands of Hollywood was not considered to be in the 'public interest' (White, 1990, p.147), but particularly useful to the FCC was the outcome of the Paramount Case in 1949 when the US Supreme Court found the major film companies guilty of antitrust violations in their monopoly of production, distribution and exhibition. As Timothy R. White explains, this offered the:

Commission and the radio interests ... the perfect weapon with which to keep the film industry out of broadcasting. Because the studios were required to separate production and distribution from exhibition, the FCC argued that the ownership of television stations would restore such a relationship

(1990, p.146).

Hilmes describes how 1927 marked the climax of the film industry's attempts to move into broadcasting, but by this time radio was taking on the structural and economic features that were to characterize the broadcasting industry for decades to come (Hilmes, 1990, p.46). This was best exemplified by the increasing dominance of the major networks; but while the major film companies may have found themselves outside of radio networking and ownership, their stars and films represented very valuable assets for radio *programming* given their established appeal, popularity and glamour (Jewell, 1984, p.125). While the opportunity to gain a controlling stake in the medium diminished, Hollywood's involvement in radio programming became crucial in the 1930s and 1940s, contributing to the development of forms of such the variety special, the dramatic series (featuring big-name stars), the publicity-gossip show and the radio adaptation of successful films (Hilmes, 1990, p.63). The involvement of the Hollywood film companies in broadcast production at this time is complex. In the late 1920s, companies such as Warner Bros, Paramount and MGM developed their own radio shows, although these did not appear on the established networks. In other cases, film companies entered into partnership with networks (such as the RKO-NBC programme *Hollywood on the Air* which began in 1932) (Hilmes, 1990, p.57). In many of the big network shows, however, control remained with the advertising agency and its client (such as *The Lux Radio Theater of the Air* beginning in 1934), with Hollywood stars and properties being contracted and hired (Ibid, p.69). For this type of show, film stars were paid well for their services. This indicates the rapid formalisation of radio's relations with Hollywood, as well as its keen demand for its assets.

As in the British context, then, it was in fact the *coverage* of film culture that was central to Hollywood's relations with broadcasting from an early stage. Yet this outline also indicates how the extent of this participation, and its sheer scale, was less feasible in the British context. It would have been impossible, for example, for film companies to produce their *own* film programmes for radio (even if they had the resources). Buscombe's comment that the 'structure of television in the UK made it more difficult [for film companies] to get involved at the production level' (1991, p.206) is equally true of radio although, prefiguring television, the film industry *participated* in film-related programming on BBC radio from the late 1920s. While important to acknowledge how the commercial nature of American radio shaped the textual form of film programming in ways which differed from the British context; BBC radio was certainly a site for symbiosis with film culture - offering coverage of films, stars and adaptations on a regular basis. In this respect, Anderson's comment that Hollywood cinema and radio were 'perceived as complimentary experiences in which stars and stories passed easily from one medium to another' (1994, p.14), is also applicable to the British situation. This also emphasises how the history of media reception and experience exceeds economic and institutional concerns.

In terms of television, its growth in Britain was considerably slower than the US. This also meant that, in comparative terms, television's inroads into cinema admissions were more gradual and less dramatic. In the UK, 1946 was the cinema's peak year, followed by a gradual decline over the next ten years. In 1950, 30 million people attended the cinema on a weekly basis, but this had fallen to ten million by 1960 - a drop of 66% (Stafford, 2001, p.96). But as Stafford notes, the rate of decline was uneven throughout this time, and prior to the advent of British television's second channel (ITV) and the wider growth of set ownership, the first half of the 1950s saw a noticeably less dramatic decline in cinema admissions. Up until mid-decade, he claims that the industry could 'still look forward optimistically, hoping for stability at audience levels that were at the average levels of the 1930s' (Stafford, 2001, p.97). This perspective was not entirely unreasonable given that attendance figures in the late 1940s represented a peak encouraged by the exceptional conditions of wartime. What is important here is that, certainly in the earlier part of the decade, there was no clear conception of what television 'meant' for the future of the cinema, and the British film industry's approaches to television were particularly multifaceted and contradictory.

One approach was again to develop forms of economic and technological *control* over broadcasting, a strategy which, although in differing degrees, characterised early approaches to television in both national contexts. In the US, the Hollywood majors investigated alternatives to broadcast television in the form of Theater Television (with television beamed into cinema theatres) and Subscription Television (a pay-per-view system which sought to bring the 'box office to the home'). Despite the fact that comparisons have rarely been drawn, British film interests also attempted to develop their own intervention in big-screen television in the form of 'Cinema-TV' (Buscombe, 1991, Macnab, 1993, Stokes, 1999). Although the idea emerged in the 1930s, it was not until the late 1940s that discussions between the BBC and the Rank Organisation became more earnest. Along with the Associated British Picture Corporation (ABPC), the Rank Organisation's stake in all areas of production, distribution and exhibition made it a vertically integrated company which competed on the closest terms with the Hollywood majors. It was such companies that were in the best position to experiment with the exploitation of broadcast technologies. ABPC, as well as the Granada cinema chain, later followed the Rank Organisation in expressing an interest in Cinema-TV. Yet as Buscombe explains, Sir J. Arthur Rank never made it entirely clear to the BBC what he wanted to do with Cinema-TV, whether he intended to transmit live coverage of events to paying audiences, offer a full range of television programming, or simply distribute cinema films to theatres via electronic means (1991, p.199). The BBC clearly perceived such plans as a threat to their monopoly, and were suspicious of the idea from the start. In any event, the film interests were never granted a license to broadcast. In its evidence to the Beveridge committee (1950) the BBC reaffirmed the view that the 'future of television would lie in the home' (Buscombe, 1991, p.201), but the film industry did not immediately give up hope. While *Current Release* was on air the trade papers were littered with news on the development of Cinema-TV and in 1953, Rank and ABPC used theatrical exhibition to transmit their films of the Coronation. Theater Television was already running into trouble in the US by this time (not least of all because of problems with frequencies, the expense of equipping cinemas CinemaScope technology, and the general expansion of television ownership) (Hilmes, 1990, p.120), and by 1953, its wider growth had not materialised. However, partly because of the slower dissemination of television in the UK it is clear that, in the early 1950s, the British film industry's forays into Cinema-TV indicated their belief that the institutional and technological future of television was still uncertain. This emphasis on a 'public', theatrical location for television, particularly when placed alongside the fact that the cinema was initially envisaged as a *domestic* technology which would bring 'the world into the home' (see Kramer, 1996), tends to undermine essentialist conceptions of the technological and aesthetic identities of cinema and television. As Hill and McLoone describe, 'the manner in which they have been developed and adapted reflects the strategic and economic needs of the respective industries rather than the aesthetic potential or limitations of either' (1996, p.3).

Although not fully coming to fruition, these interventions into Theater and Cinema-TV emphasise how, throughout the history of cinema's relations with radio and television, it is largely cinema *exhibitors* who have had the most to lose from media competition (Hilmes, 1990, p.42). While for producers and distributors, radio and television have historically represented further exhibition outlets for their products, exhibitors remain attached to their theatres. It seems significant here that Granada, Rank and ABPC - the three British film companies to chase the possibilities of Cinema-TV - all owned cinema chains. (Rank owned the Odeon and Gaumont circuits, and ABPC operated the ABC chain). Exhibitors backed Cinema-TV because it was clearly not a form which cut the cinema out of the equation, and it also represented little threat to film producers. This emphasis on the different fractions of the film industry also explains why and how it was feasible for contradictory approaches toward television to be in evidence simultaneously. While the debates over Cinema-TV reached their climax, the film industry were in constant negotiation with the BBC over publicising their films on television via *Current Release*, while at the same time stepping up their refusal to sell any feature films. It is certainly the exhibitors who were often at the forefront of negotiations over the cinema programme in the 1950s, and as Chapter Three will demonstrate, fearing that it might usurp the audience for 'film' as much as promote the cinema, they increasingly stipulated *when* the programmes could be scheduled. This in itself often created an interesting disparity between the reports on the programmes in the trade press, and the BBC's own correspondence with individual film producers and companies.

Despite these various attempts to move into broadcasting, by the mid-1950s neither the Hollywood majors nor the British film industry had any significant ownership interests in television in terms of transmission or technology, although the attempts by the Hollywood industry to appropriate broadcasting had occurred on a much larger scale. Nevertheless, in both national contexts, it was in fact at the level of film-related programming (particularly when we look back to radio), that some of the most culturally visible forms of interaction between cinema and broadcasting had occurred - even if the scale, form and structures which fuelled this coverage differed. In the mid-1950s, however, there are significant shifts in the industrial and institutional organisation of television in both Britain and the US, with significant implications for shaping relations with the film industry. These shifts again point to certain similarities between the contexts which are often overlooked, but they also emphasise the differences in the situations, contextualising the role played by the cinema programme on British television at this time. In short, what we see as a result of this situation is that the significance of film-related programming in approaching the historical relations between Hollywood cinema and television declines. In the British context, however, it emerges as the primary site of interaction, with a cultural and institutional centrality it has not enjoyed since.

Although the American majors were previously prevented from moving into television production, a number of factors reshaped the market for television programming in the 1950s, with considerable implications for the film industry. American television experienced a considerable shift in its economic and institutional infrastructure. In the early 1950s, the system of programme production established by radio continued, and the control over programming still rested with the commercial sponsor (often via an advertising agency) (Hilmes, 1990, p.140). But as Hilmes describes, things were about to change: 'from a medium controlled by the interests of its sponsors ... network television became a medium controlled and programmed by the major networks, with Hollywood as its primary programme supplier' (1990, p.137). By 1957, a new structure was in evidence which placed the increasing control of *production* with the major television networks. As with commercial television in the UK, programmes were now interspersed by advertisements, and the networks sold time slots to the advertisers (Ibid, p.140). This changed infrastructure helped stimulate demand for programming which created a lucrative opportunity for Hollywood. With the decline of live programming and the increasing use of filmed material, the 1950s saw the rise of the telefilm series, and following the earlier example of the smaller film company Columbia (responsible for TV series in the early 1950s such as *Father Knows Best*), the majors had all moved into television production by 1955. Unlike the earlier independent producers of telefilm series, they supplied programming to the networks during prime time, and rapidly colonised the market (Balio, 1990, p.32). With the potential for syndication and foreign distribution, this method of production also enabled the majors to capitalise on aspects of the studio system by shifting mass production into series TV (Anderson, 1991).

The majors' move into television production, however, was initially prompted less by the lucrative appeal of the telefilm series than by the opportunity to use television to promote their current films. Walt Disney Productions led the way when they signed with the ABC network to produce *Disneyland*. This was a weekly series centred around the company's theme park, but with the added right to include a six-minute segment which promoted Disney's current films (Balio, 1990, p.33). As Balio describes, 'It was this concept of using programming as both product and promotion that the studios found alluring' (Ibid). Hollywood films and stars already received considerable coverage on American television in variety programmes such as *The Ed Sullivan Show*, and they clearly continued to do so as the 1950s progressed (Anderson, 1994, 160). However, in order to take control of the 'previously pervasive but scattered use of Hollywood film stars and general glamour on television' (Hilmes, 1990, p.152), in 1955 Warner Bros, Twentieth Century Fox and MGM all diversified into television with flagship series carrying the company name: *Warner Bros Presents* (1955-57), *Twentieth Century Fox Hour* (1955-6) and *MGM Parade* (1955-6). It is worth noting here, however, that unlike the British *Current Release*, *Picture Parade* and *Film Fanfare*, these were not 'cinema programmes' *per se*. While this could conceivably be seen as forcing a retrospective conception of the genre onto these early years (the argument that they aren't recognisable as cinema programmes from a contemporary perspective so they don't 'fit'), it is at least possible to suggest that they are not primarily organised around the discussion and coverage of contemporary film culture. The Warner Bros programme was an episodic series based on adaptations of the company's successful films, but in 'exchange' for this, Warner's were able to include a six- to eight-minute segment entitled 'Behind the Cameras at Warner Bros' which included promotion of the studio's current films. Significant here is the extent to which scholars are united in emphasising the relative failure of these series and consequently, their more marginal significance in approaching the early relations between Hollywood and television (Balio, 1990, Anderson, 1994, Hilmes, 1990). As Hilmes reminds us:

> However, these self-promotion vehicles, although indicative of Hollywood's continu-
> ing interest in and recognition of the importance of television, proved to have far less
> effect in the long run on the television broadcasting industry - and on the film indus-
> try - than did production of the less glamorous but more profitable filmed series with
> which Hollywood slowly came to dominate broadcast schedules

(1990, p.155-6).

Representing a loss for the studios and apparently unpopular with audiences, both networks and advertisers complained about the perceived fragmented and unbalanced nature of the shows - not least of all their 'overcommercialisation' and blatant 'plugs' for the films (Anderson, 1994, p.197). For various reasons they were short-lived, prompting the majors to concentrate on the more routine yet profitable field of telefilm production.

What is most significant here is that the organisation and development, as well as the textual forms, fostered by the US film and television industries led to a situation in which the concept of a cinema programme did not have a clear institutional or cultural identity at this time. In terms of the tensions encountered above, the *promotional* implications of the British cinema programme were undoubtedly also a site of constant debate (particularly in the context of public service and restrictions surrounding 'advertising'). Yet in America, with the studios producing their *own* footage to publicise their *own* films, it would appear that what Hilmes has elsewhere described as the 'delicate balance' (1990, p.93) of the interests of broadcaster and film studio was difficult to maintain. Furthermore, these shows were representative of a different type of text and were not in any case devoted to the coverage of film - even if this element was the film companies' primary motive for their production. Hence, while the changing trajectory of Hollywood's relations with television in the 1950s clearly leads us back to Anderson's statement that 'history of American TV is the history of Hollywood TV' (1994, p.12), this refers to the move into television production on a broader scale.

The relations between the film industry and television in Britain also changed in the mid-1950s as a result of the introduction of commercial television (ITV), a second channel which represented a challenge to the BBC's monopoly. The new channel was constructed by a network of regional stations funded by advertising - all of which were regulated by the central authority of the Independent Television Authority (ITA) (Stokes, 1999, p.33). Particularly in terms of its implications for the BBC, the emergence of ITV is discussed in Chapter Four, but as Stokes explains, 'Commercial television promised greater choice of viewing, an expansion of the market, and a new optimism about popular broadcasting' (Ibid). Significant here are the implications of ITV for the broader relations between British television and the film industry. What commercial television offered the film industry was the chance to finally gain a foothold in the medium at an institutional level (in such a way that was not in fact available to the Hollywood film companies in the 1950s). British film companies ABPC and Granada submitted successful bids for ITV network franchises, becoming programme contractors in 1956. Although not directly involved in the running of ITV, the Rank Organisation also purchased shares in Southern Television the following year (Paulu, 1961, p.168) and, as discussed, all three companies exhibited an interest in television from an early stage in their attempts to develop and operate Cinema-TV. In her brief overview of the relations between the British film industry and television in this period, Stokes argues that the:

> attitude of the ITV companies toward the film industry was entirely different from that of the BBC, and the relationship between the two sectors was transformed by commercial television. From its inception, commercial television in Britain had the backing of cinema interests

(1999, p.33).

ITV instigated an important shift in the institutional relations between the British film industry and television, but as Stokes' argument indicates, it is easier to see this as representing something 'entirely different', or as a 'transformation', if it is assumed that the BBC had virtually nothing to do with the film industry at all.

The dynamics of ABPC's move into television (as ABC TV) will be discussed in Chapter Four, but it is clear that this shift generated much discussion in the film industry's trade press, and the film companies concerned were well aware that their move into ITV was controversial. In the face of increasing competition from the medium, some commentaries saw the move as a shrewd and practical decision, while others perceived it to be the ultimate betrayal, insisting that such companies could not continue to serve 'both camps'. As a result, it is perhaps not surprising that the Managing Director of ABC TV, Howard Thomas, was repeatedly quoted in the trade press as saying that ABPC had 'gone into television to boost cinema admissions' - apparently referring to the possibilities of programmes that would promote the cinema. Given the huge task of running an ITV franchise and the fact that (in terms of total broadcasting hours), the proportion of time that could be devoted to film-related programming was comparatively small, it is difficult to believe that this was the primary reason for the creation of ABC TV. Such comments seemed to be as much intended to appease the complaints and fears of the film industry while playing down the accusations that ABPC had moved into the 'enemy camp'. Intriguingly, however, Thomas' comment here recalls the motives of the Hollywood majors in terms of their initial move into television production (with their promotional shows), and when ABC TV began to broadcast in February 1956, their flagship cinema series, *Film Fanfare*, was one of the first programmes on air. *Film Fanfare* was a site upon which the various interests of ABPC - television, film production, distribution and exhibition - coalesced in multiple ways to the economic benefit of the parent company. However, in terms of a comparison with the short-lived series in the American context, this could not be conceived as essentially a cinema programme produced by, and promoting, one company's films (which would undoubtedly have contravened the tight regulation of 'advertising' on British television). While the links to ABPC and film production/ exhibition certainly permeated the series, it was essentially a cinema programme, comprised of the much broader sweep of film culture that this implies. Although film interests had moved into television and, in certain spheres, could use this to their advantage where the cinema was concerned, they were first and foremost to operate as television programme contractors, responsible for providing the general output of ITV. As in Hollywood, this changed structure, the wider expansion of television, and the decline of live programming, saw the rising importance of the telefilm series on British television and with it, the provision of work for the film industry. This shift also provided work for independent film companies, although it is important to stress that the nature of both the British film and television industries meant that this was by no means approaching anything like the scale of TV production in Hollywood at this time.

While we can talk of the film industry 'moving into' British television at this time, this is in a different way to the US, and with different implications. In terms of contextualising the centrality of the cinema programme in the early relations between British television and film culture, it can certainly be seen that, although the BBC continued to play an equally important role, the introduction of ITV marked the beginning of a period of intense activity where this coverage was concerned. Although the fact that certain film companies had moved into ITV may have increased the scope and energy of the coverage the channel offered to the cinema, it was as much to do with a second channel increasing the opportunity for such programming on a wider scale. Within this context, and arguably because the institutional organisation of British television and cinema could not offer the opportunities for the wider interaction seen in the US, the cinema programme continued to function as a crucial point of institutional and cultural dialogue and negotiation between the media. In discussing the cinema programme in the mid-1950s as continuing to function as a form of 'negotiation' (as well as a site of developing interrelations) between the film industry and television, this term primarily refers to the fact that television still had difficulties in obtaining feature films. A key argument of the book is that, contrary to prevalent perceptions at the time, the cinema programme did not primarily represent a promotional bargaining tool, a 'free' shop window for the film industry with which television hoped to obtain feature films in return. However, to a certain extent, these histories are interlinked, and the cinema programme was viewed as a site upon which to foster wider working relations with the film industry from an early stage.

The experiments with Cinema-TV emphasise how film interests were keen not to increase the *domestic* attraction of television, and despite their willingness to enter into the possibilities of a cinema programme, their refusal to supply the medium with feature films was not unsurprising. Although British companies such as Rank, Ealing and London Films had sold films to American television in the late 1940s (Buscombe, 1991, p.203) (largely screened at fringe hours on independent stations), the British film industry remained as opposed to selling their wares to British television as the Hollywood majors remained reluctant to release their features to the US networks (see Lafferty, 1990). Yet it is worth noting here that the histories charting the sale of feature films to television have primarily focused on the 'facts' concerning the institutional negotiations and the junctures when films were obtained. This has often elided a considered of the cultural dimensions of these relations. What has not been explored is how such films were contextualised for the viewer, how and why particular types of channel screened particular types of film, and the ways in which they were received (Holmes, 1998). These dimensions are largely outside the scope of this study, but are touched on briefly here.

'Regular films in the comfort of your home': Feature Films on Television

It is clear that the amount that the BBC were willing (or able) to pay for the films was often a key reason for negotiations breaking down, and their correspondence with the film industry confirms Buscombe's point that they 'could not afford to pay much' (1991, p.202). In 1952 the British Cinema Exhibitors Association (CEA) attempted to formalise the film industry's position on the matter and passed its 'Llandudno Resolution': to discontinue trading with any renter or producer known to supply feature films to television. As a result of this embargo, virtually all the films the BBC could obtain in the pre- and immediate post-war periods were those in a foreign language (Buscombe, p.198). However, contextualisation in the *Radio Times*[1] seems to suggest that well-known 'classics' such as *Les Enfants Du Paradis* (Marcel Carne, 1945), *Paisa* (Roberto Rossellini, 1946) and *Bicycle Thieves* (Vittorio de Sica, 1948) were presented and scheduled as very much 'prestige' features, with the Corporation defining them as 'high culture' and capitalising on the renaissance in European cinema at this time (Holmes, 1998). Aside from this, in the late 1940s and early 1950s the BBC had to make do with a very limited selection of older material, and certain British and American comedies from the late 1930s were occasionally screened during the day. In the mid-1950s the BBC (as well as ITV) still screened very few films, and those that did appear were offered as a specific holiday or seasonal 'treat'.

A survey of the *Radio* and *TV Times* indicates how it is not until the period from 1957 to '58 that there emerges a shift in the screening of feature films on British television. The ITV contractor ATV acquired an extensive library of British films (1939-46) from the independent film company British National, which were gradually moved to a fortnightly slot on Sunday evenings (10:05pm) and notably identified as the 'ATV Late-night Movie'. A further shift occurred in September 1957 when, following the death of Alexander Korda, 25 of his films passed to ITV (such as *The Private Life of Henry 8th* (1933) and *Catherine the Great* (1934)). Although still from the 1930s, these films were understood and promoted as 'special' season - with the front cover of the *TV Times* promising 'Great British Films... The Korda Season Starts Here'.[2] However, although screened on Saturday nights, in an effort to appease complaints from the film industry the films were still broadcast just outside of peak hours at 10:05pm. In 1958 the BBC acquired 100 films from RKO (ranging from pre-war productions to some as recent as 1955), and to the horror of the film industry, they were screened every other Saturday, alternating with *Saturday Theatre*, at the peak time of 9:00pm. ITV then sought to compete and acquired a substantial number of Warner Bros films (mainly produced between 1936 and '48 and often starring Bette Davis, Humphrey Bogart or Edward G. Robinson) and screened them on Saturday nights at 8:30pm. The age of feature films on television had begun, as announced with considerable fanfare in the *TV Times*:

Regular films in the comfort of your home. Not just any films [but] GREAT films, top entertainment cinema classics. They will open the eyes of those who haven't seen them and revive great movie-going experiences for those who have, so wait for the curtain to open and prepare for what is in store... All these and more - for as the cinema advertisements have so often told us - your future entertainment.[3]

Invoking precisely what the film industry had most feared, this presented film on television in the terms of a 'home-movie theatre', a space which claimed to approximate the service of the cinema itself. The same year the film fan magazine *Picturegoer* began its 'Home Screen Film Parade', announcing that: 'Because weekend stay-at-home picturegoing is looming larger and larger, [we] ... are providing a weekly commentary on the films being shown'.[4] (In comparison, although the Hollywood majors began to sell their pre-1948 libraries to American television by the mid-1950s, it was not until 1961 that feature films were scheduled as a prime-time offering on the networks) (see Lafferty, 1990). With increasing broadcasting hours and audiences, however, British films were still an important prospect for BBC and ITV given the restrictions on the amount of 'foreign' (including American) material they were allowed to screen. It was in response to the feature film's increasing penetration of the television schedules that the Film Industry Defense Organisation (FIDO) was formed in 1958 which was intended to prevent the sale of feature films to television. It was to do this by placing a levy on cinema admissions (one farthing per seat), generating funds which FIDO would then pay to producers as compensation for the money they would have received had they sold their films to television (Buscombe, 1991, p.203). As Buscombe notes, it was perhaps a sign of 'desperation that such a fragmented industry unite around such an unlikely scheme' (Ibid), particularly as it emerged so late in the 1950s when the economic and cultural prevalence of television seemed increasingly undeniable. Very much a short-term strategy to try and limit television's increasing popularity, FIDO disbanded in 1964 due to lack of funds. As Buscombe observes, it seemed like a strategy which was 'doomed from the start. It took funds from a constantly declining revenue base, the cinema box-office, and used them to compete in the market with a rival whose economic strength was increasing every year that went by' (Buscombe, 1991, p.205).

Aside from the institutional disputes and negotiations, this context is also important in considering issues of reception at this time as it emphasises how, for much of the decade, the cinema programme represented the primary experience of 'film' on television. It was precisely within the context outlined above, the limits which structured the wider institutional relations between British film culture and television and the relative absence of full-length films, that the cinema programme flourished. Indeed, as the end of the book makes clear, it is at the very point when the age of feature films on television becomes conventionalised that the cinema programme seems to lose its centrality on a number of different levels.

'An Ever-Increasing Amount of Leisure': The Cultural Contexts for Screen Entertainment

These changing trends in the relations between cinema and television, as well as the consumption of screen entertainment, clearly took place within a wider cultural context in the 1950s which, whether in terms of social and cultural surveys of the era (e.g. Hopkins, 1963), or broader discussion of the period in media histories (e.g. Geraghty, 2000a), is fairly well documented. As Geraghty has observed, within this context, 'declining [cinema] audiences and the reasons for this become something of an obsession in commentary about cinema in the 1950s' (2000a, p.6), and this has continued in subsequent histories of the period, particularly where cinema history is concerned. Given that this statistical evidence is limited in what it can tell us about the changing experience of cinema and television at this time, as well as the relations between the media, it is used sparingly here. The changing roles of cinema and television are discussed throughout the book and issues of audience reception are considered in detail in Chapter Three.

While television is often perceived as the primary causal factor in the cinema's decline, their fortunes were also shaped by wider social and cultural contexts. A number of factors worked to diminish the cinema's importance and increase television's appeal. As is widely acknowledged, in terms of economic prosperity, the beginning of the decade saw a continuation of the austere conditions of wartime, but affluence quickly became what John Hill describes as a 'central economic, as well as ideological theme of the times' (1986, p.9). The effects of affluence reached Britain later than the US, and 1953 has been cited as a significant point of change (Stacey, 1994, p.222). The combination of new techniques of mass production with newly available forms of credit produced an expanding consumer market and a concurrent rise in the standard of living. Studies of British domestic leisure in the late 1950s revealed that the emphasis on do-it-yourself, decorating and gardening ensured that 'more money is used for the house, and more leisure is used for work' on it (cited in Geraghty, 2000a, p.12) - in itself intrinsically linked to, and evidence of, greater suburbanisation. Television was one several activities based around the home, detracting from time spent on the cinema and other forms of public entertainment. For example, a *Picture Post* article from 1956 on the declining pub trade self-consciously observed how 'social patterns are shifting ... each year new millions become Televiewers... It's a sip of bottled beer around the tele-screen at home against a night out in the saloon bar'. [5]

Contrary to the belief that the film industry remained fixated on the impact of television as other factors remained more intangible and elusive, the trade often recognised a number of contributing causes where the decline in cinema attendance was concerned. Following an audience survey in 1956, *Kine Weekly* noted that:

Among a variety of causes to which this present situation has been ascribed, television... is among the most constant. This has never, in fact, been proved, but the general public has been convinced by newspaper stories that this is so.[6]

Insisting that 'all the industry's present troubles could not be laid at the door of television', the trade paper had earlier recognised 'affluence' as a contributing factor when it noted that it was not simply television but 'better housing conditions' that were accelerating the decline in attendance. The critic discussed how patrons used to attend the cinema to escape 'the dull and unpleasant' home and environment and asked, 'if some of the old reasons for going to the cinema no longer apply, what is the solution to the problem?'[7] Yet despite (and in some ways because of) the harsh conditions endured in wartime, the late 1940s was a boom period for British leisure industries more generally, and the 1950s was an acceleration of this trend, rather than a radical shift (Walvin, 1978, p.149). Many of these pursuits, rather than contributing to an increased emphasis on the domestic sphere, in fact took people *away* from the home (see Hopkins, 1963), not least of all because of the gradual increase in car ownership. As a critic in the *Daily Film Renter* observed, the cinema was competing not simply 'for people's money, but for their time' and quoted a Hollywood producer who suggested that 'the potential audience consists of a constantly growing population having at its disposal an ever-increasing amount of leisure'.[8]

Nevertheless, some of these key shifts support what is known as the privatisation thesis. The idea that family life has become increasingly privatised since the post-war period is an influential one, and as Graham Allan and Graham Crow explain, various terms are used to describe these changes such as 'house-centered existence' or 'home-centered society' (1991, p.19). Although privatisation is often used as an all-encompassing 'umbrella term' which can obscure the complexity of the factors involved (Allan and Crow, 1991, p.19), Allan and Crow rightly point out that 'we need to be... skeptical of the stereotypical model which suggests that at some time past, social life was more communal and less focused on the household... than it currently is' (p.24). There is certainly a danger in discussing the post-war period as though there was a sudden 'retreat' to the home. Furthermore, privatisation is often constructed as an increasing trend since the post-war period, and it is unclear how relevant it is in understanding the more immediate post-war years, and perhaps more importantly, how it was actively *experienced*.

The cinema programme is clear evidence (and was part) of the extent to which the cinema's identity as a popular mass entertainment continued well into the 1950s. Nevertheless, this is a period in which the cinema clearly begins to loose its wider appeal - in which context its cultural associations, experience as a form of public space, and role in everyday life, begins to change (Geraghty, 2000a). The 16-24 age group had always represented the majority of the cinema audience, but between 1946 and 1960 the proportion of this group nearly doubled in size (Harper and Porter, 1999, p.67). Towards the end of the decade, the cinema becomes increasingly associated with and frequented by 'teenagers', a term coined in this period to describe the youth market with an increasing amount of disposable income. By the end of the 1950s the concept of the family audience enjoying a regular night out at the cinema was no longer the primary way that the cultural role of the medium was conceived (Geraghty, 2000a). Although the changing social and cultural status of the cinema is contextualised in more detail in the final chapter, Geraghty explains how:

> The cinema was thought of and presented itself as old-fashioned, uncomfortable and associated with past pleasures. For the general audience, cinemagoing was changing from being the quintessential modern form of popular entertainment to an old-fashioned and somewhat marginal pursuit
>
> *(2000a, p.20)*.

It is within this changing context for screen entertainment that the cinema programme develops, flourishes and indeed declines. It offers a different perspective on the narrative of these shifts, and the ways in which they can be approached and understood.

Hence, this chapter has emphasised the economic, institutional and cultural factors which shaped a particular space for the cinema programme on 1950s British television, and its role in British film and television culture. In indicating the wider contexts of the relations between the media at this time, it is also significant to note that more attention has been paid to spheres of interaction which were effectively 'non-starters' - the attempts to develop Cinema-TV, and the endless (and for much of the decade, fruitless) disputes over the sale of the feature films to television. In approaching these relations, broadcast programming, whether conceptually, historically or literally, is typically marked as ephemeral and transient, and, when compared to the well-documented industrial and institutional encounters, is perhaps not as easy to 'pin down.' It is to reconstruction of its world we now turn.

Notes
1. See for example C. A. Lejeune's contextualisation of *Les Enfants Du Paradis* in the *Radio Times*, 30 December, 1949, p.21.

2. *TV Times*, 21st September, 1957.

3. *TV Times*, 19 January, 1958, p.3.

4. *Picturegoer*, 1 February, 1958, p.13.

5. *Picture Post*, 28 April, 1956, p.40.

6. *Kine Weekly*, 13 December, 1956, p.8.

7. *Kine Weekly*, 5 July, 1956, p.28.

8. *Daily Film Renter*, 27 September, 1956, p.2.

Chapter Two

The Cinema Programme Begins:
Developing Film 'Specially For Television Purposes and Technique'

This chapter traces the development of the televised cinema programme from its status as a concept in the late 1930s, to just before its appearance on air in 1952. While the televised cinema programme did not actually come to fruition in the 1930s, this period is revealing in suggesting how the 'idea' of the genre developed, and the institutional, economic and technological factors which shaped this process. This includes in particular the cinema programme's relations with public service, and the ways in which the genre, and the interests of the film industry, generated a number of tensions within this sphere. As the history of the BBC's relations with film culture on radio also represented an important context for the development of a televised cinema programme, the historical parameters of this are considered here.

Critical Perspectives on the Cinema Programme

Cinema programmes in the 1950s were consistently held in low critical regard, although this is not a perception specific to this period. While concerted consideration of the genre remains scarce, television has often been accused of failing to develop a sufficiently analytical, challenging and inquiring approach to the cinema. Ed Buscombe's *Screen* pamphlet *Films on TV* (1971) (considering both cinema programmes and films on television more generally), represents one of few academic analyses of the genre. Commenting on the BBC's now long-running *Film...* series (then still relatively new and presented by Peter Jenkins), Buscombe described its attitude towards the cinema as superficial, 'passive' and formulaic (1971, p.45). It failed, he claimed, to take a 'serious' perspective on film and was simply 'news and views' rather than an exploration of 'ideas about the cinema' (p.41). The function of the pamphlet was not simply to provide a critical appraisal of the genre, but also to recommend strategies for improvement. In this respect, Buscombe argued that television should 'redefine its relationship to the cinema' and concentrate 'more on ideas, and less on ephemera and trivia... [Television] has a responsibility to do more than use the cinema as a source of cheap programmes and second-hand show-biz glamour' (p.49/p.60). Implicit in the reference to 'responsibility' is that television is failing to fulfill the central tenets of its public service remit - to use the arts to educate and inform, as well as entertain.

Buscombe was considering the programmes at a time when the serious study of television was only just emerging (and even the development of film studies was still in its early stages), and this may have shaped his unwillingness to offer a more sympathetic perspective on the programmes. His perception of the genre is certainly similar to opinions expressed in the popular press at that time. The *Daily Express* described the programmes as 'prisoners of commercial interests' which exhibited a 'bland' and 'benign' attitude toward film. The critic lamented the 'hoary old "clips and chat" format which has straitjacketed the genre'[1] and argued that the programmes should concern themselves with more complex issues such as the relations between films and the social context from which they emerge. The *Daily Express* concluded that 'The level of TV's approach to the cinema can by no tolerant stretch... reach that of similar arts shows on radio', and cited such programmes as examples of 'intelligent and acute criticism'. To take an example from later in the decade, film academic Colin McArthur attacked the televised cinema programme in his film column in the *Tribune*. In 1978, while he applauded the BBC's *Arena: Cinema* for the 'knowledge' and 'commitment' with which it approached its subject, features such as *Film '78* and *Hollywood Greats* (a star-profile series), are criticised for their 'ignorant', 'trivialising' and superficial attitude.[2]

Given that the cinema was still more of a mass medium in the 1950s we might expect critics then to have different expectations of television's coverage of film, but this was not necessarily the case. Although they were perhaps less inclined to openly demand a more 'serious' approach to the cinema, they still considered the programmes frivolous, superficial and shallow. These tensions predominantly revolved around television's freedom to *criticise* cinema - an issue pivotal in shaping the responses of Buscombe and others in later years. In the 1950s, however, these criticisms of the genre were shaped by perceptions of hostility where the cinema's approach to television was concerned. Not unlike the academic work on the early relations between British cinema and television, the press were keen to play up the antagonisms regarding the sale of feature films to television, and this was crucial in influencing the reception of the cinema programme and perceptions of its cultural value. With respect to the BBC's first cinema programme, *Current Release*, the *Daily Mirror* insisted that the Corporation should 'Take this film "plug" off! ... Television introduced this feature hoping the film world would give them films in return. This programme has failed to prise any more celluloid from filmland [so] now is the time to take it off!'.[3] The *Daily Star* agreed that *Current Release* 'only reached our screens in the first place as bait to the film people'[4] and the *Daily Express* asserted that 'As a sign of good faith, TV has given the film industry free publicity worth many thousands of pounds... The BBC is so anxious to help that... the film [industry's] own advertising staff could not have done better'.[5] Significantly, critics of the time also presumed to speak for the reception of *Current Release*. The *Daily Mirror* simply stated that 'This programme is pretty poor value for the viewer' while the *Star* insisted that it was 'a perpetual reminder of what TV might have but never gets'- that is, feature-length films. In reading the available press reports we could be forgiven for thinking that these programmes were an utter embarrassment to the BBC (and later ITV), and that they were disastrously unpopular with viewers - although this is far from the truth. Nevertheless, what is important here is that the status of these early responses as 'evidence' has also shaped subsequent perceptions of the genre. In 1956 Burton Paulu's book *British Broadcasting* similarly implied that they were nothing but a 'bargaining' counter in the bid for feature films and simultaneously, an attempted substitute for their absence. Over thirty years later, Ed Buscombe cites Paulu's work as demonstrating how: 'So eager was television to get... a taste of film industry glamour that it was willing to countenance programmes that were little more than shameless plugs for new Hollywood releases' (Buscombe, 1991, p.202).

Pre-war Television and the Cinema Programme

In re-examining the evidence on which these accounts are based, the narrative of the cinema programme begins in 1936. In the 1930s the cinema was the most popular mass entertainment medium and it is thus not surprising that before the television service opened in 1936, the BBC approached several film companies regarding the availability of feature films for television. As the range of the television service was very limited at this time (and there were only a small number of sets in use), the BBC assured the film industry that 'No damage to [the cinema] is even remotely possible', and they clearly envisaged a co-operative co-existence.[6] With this peaceful partnership in mind, the Director General of television, Gerald Cock, explained to the film companies that the BBC would not screen any film more than three times and that they could come to an arrangement by which 'a film would not be televised until after its London release'.[7] It is evident that, from the BBC's point of view, television's threat to the cinema was perceived to be so small that even current films were not beyond its reach. Yet it was not only through the full-length feature that the BBC intended to bring contemporary film culture into the home. With the opening of the television service they intended to transmit two cinema programmes per week which dealt with films on West End and general release. The BBC also hoped to produce 'occasional programmes, such as the tracing of a "star's" or director's career by means of brief "flashes" from a number of their pictures'.[8]

In challenging the assumption that the cinema programme was simply a 'bargaining' tool in the bid for feature films, this series was solely the conception of the BBC and was evidently suggested *before* the companies refused to sell their films. Cock's proposal was also shaped by the BBC's prior experience with film on radio, and at the time of Cock's proposal, weekly film reviews were routine. Film, book and theatre reviews were all given similar treatment and were known as 'service talks'. The term 'service' here was used in the context of 'public service' and hence directly invoked the function of the reviews. As the Director of Talks explained, the film reviews were intended to help the 'interested listener' select the features they wished to see.[9] To a certain extent, therefore, Cock's proposal was simply an extension of radio and the general broadcasting policy of public service.

Both culturally and institutionally, television was often perceived as largely an 'extension' of radio until at least the early 1950s (Paulu, 1961, p.54). However, there were also significant differences. The service talks on radio did not simply inform the listener of the films on release but gave 'a critic's opinion of them'.[10] This was in contrast to the television programme which proposed to

> deal with films descriptively (rather than critically).... The speaker's remarks would refer chiefly to interesting facts about the stars, production, direction, and story, and would be illustrated by 'floating in'... visually and in sound, one or more brief excerpts from each of the films concerned.[11]

The BBC had often conflicted with the film industry over freedom of criticism where the film reviews on radio were concerned. Companies had readily complained when their films attracted adverse comment, and Cock was clearly mindful of these tensions when proposing the television programme. However, his notion of a 'descriptive' rather than 'critical' feature was perhaps shaped by factors other than a desire for the industry's co-operation. He may have understood that in order to make the feature an attractive prospect for television, a simple adaptation from radio would be inadequate. The public, of course, could obtain a straightforward review via sound. The 'interesting facts' about the stars or production, complete with 'floating in' the clips concerned, played a role in constructing the visual appeal of the programme. As Cock elaborated: 'We should very much like to count upon the co-operation of your company in what will... turn out to be a new and interesting form of "trailer", designed specially for television purposes and technique'.[12] In using terms such as 'floating in' and 'flashes', Cock struggles to find the vocabulary to describe what is an unfamiliar terrain. However, even if he was uncertain as to how to conceptualise the programme he had in mind, he was clearly thinking along visual lines.

The proposal was entitled 'Special Television Film "Trailers"'. From this time on, the term 'trailer' always appeared in inverted commas which served to distance its commercial connotations. This was the start of ongoing institutional concerns about the ways in which the cinema programme - with its promotional implications - might conflict with the BBC's ethos of public service. That said, Cock's letter also suggests that, while the television 'trailer' would be more than mere 'advertising', he understood the promotional advantages it offered the film industry. This was clear in the initial enquiry concerning full-length films as the BBC had informed the companies that they 'should be prepared, as part of the scheme, to televise special..."trailers" of important new films'.[13] Presented as 'part of the scheme', the 'trailer' is here understood as an incentive which may make the companies more responsive to the BBC's demands, although we might also note that such material is envisaged as *separate* from the cinema programme itself. This seems to suggest that while the aesthetic form of the cinema programme was in part an enticement to encourage the film industry to sell its wares, it was not a bargaining tool *per se*. Its conception was shaped by the desire to provide a service to the viewing public, the history of which goes back to the inception of radio a decade before. Certainly, such a programme could act as a means to aid institutional relations between cinema and television, not only with respect to the future trading of feature films, but also in terms of the appearance of stars, directors and other artists. Cock explained to the companies how the programme could form 'a useful basis of co-operation between the Film Industry and Television', while adding that the BBC was 'not proposing (in other than perhaps exceptional cases), to televise feature films in any other form'.[14] In fact, this makes clear that, in contrast to the regular scheduling of the cinema programme, the BBC was not intending to make extensive use of feature-length films at all. The film companies refused to meet even these limited needs, not least due to problems over patent rights and the perceived technological limitations of television. The BBC later acknowledged that 'Film television [was] still inefficient technically' and that even if films had been obtained, they would have emphasised the limitations, rather than benefits of the medium.[15] This would seem to further suggest that, far from simply a cheap trade-off, the cinema programme was envisaged as an important - and in some ways 'natural' - form of partnership between British film culture and British television from the medium's earliest days.[16]

While film companies co-operated with the BBC by supplying clips for a television exhibition, they did not agree with Cock's proposed programme - with the exception of Alexander Korda's London Films. The most powerful company in the British film industry in the late 1930s, London Films were the first to co-operate with the BBC by later supplying a clip of *Elephant Boy* (Robert Flaherty/ Zoltan Korda, 1937) for the topical magazine programme *Picture Page*.[17] As co-director, Robert Flaherty appeared on the programme to discuss his work. This edition of *Picture Page* (1936-39, 1946-52) was transmitted on Christmas Day and it was clearly understood to be a special occasion - indicating the extent to which the BBC regarded coverage of film to be an appealing strand of the television service. The possibilities of this were further explored in March 1939, when a 'dramatic sequence' from Korda's *The Four Feathers* (Zoltan Korda, 1939) appeared on *Picture Page*, and was followed by an interview with its stars, John Clements, Ralph Richardson and June Dupré. The Director of Publicity at London Films described how 'An informal talk on ... filming in Sudan could precede the televising of the selected piece of film'.[18] This is not unlike the material which would eventually comprise the cinema programme, and there were already conceptions as to how it could function as television entertainment. As London Films acknowledged, 'It is of course understood that the whole thing should be made interesting from your point of view, and not merely a piece of publicity for the film'.[19] Representing the differing investments of the film industry and the BBC, the balancing of these demands would structure the debates over the cinema programme in its formative years. This was particularly so given that the majority of companies would not be nearly so willing to consider what we might term the aesthetic and institutional demands of television as the eager London Films.

One year after the television service began, the Corporation made a final attempt to get a regular film programme on air. This time, Cock contacted the Kinematograph Renters Society (KRS) which he hoped would speak with the film companies on the BBC's behalf. Including a review of West End and general releases, the proposal was largely the same as before. There was, however, one important difference. Somewhat curiously, Cock explicitly conceptualised the presenter as a 'critic' and informed the KRS that they 'must necessarily have the normal freedom of criticism'.[20] (Cock also stressed how television had improved technically and thus perhaps assumed that it was also the *aesthetics* of the medium which had previously deterred the companies from co-operating). However, in replacing the 'speaker' with a 'critic' it is not surprising that, according to Cock, the film industry 'dismissed the proposal out of hand'.[21] Following the closure of the television service for the duration of the war, it was this very debate which opened discussions on *Current Release* in 1950. The difficulties and tensions surrounding the cinema programme persisted after the war. Although the BBC television service resumed in 1946, it did not immediately return with the popular coverage of the cinema it had been so keen to put on air.

Planning *Current Release*

When the television service began broadcasting again after the war its expansion was slow. This was partly due to the economic circumstances of post-war austerity in which the manufacture of non-essential items (including televisions) was restricted, as were the materials required to broaden the BBC's area of transmission (Stokes, 1999, pp.32-3). These factors contributed to the slow spread of television ownership and in the late 1940s, the medium was still something of a luxury. It is widely recognised that it was also the institutional dynamics of the BBC which hampered the expansion of the medium, as it was forced to play a subservient role to radio for several years. Television occupied a marginal space at the back of the *Radio Times* until well into the 1950s, and in the 1940s, television programming was merely the additional responsibility of the radio departments (and accounted for less than one-tenth of total BBC expenditure) (Crisell, 1997, p.74). In 1937 the BBC calculated that one hour of television would cost twelve times as much to produce as the costliest hour of radio and as a result, Crisell describes how the 'service resumed not so much with the expectation that it would develop as that it would need to be curbed' (Crisell, 1997, p.74). *Current Release* emerged from these early stages in television's post-war development, a time when its expansion was just beginning to gain ground. Television was given its own department in 1950, although it still only broadcast for a limited number of hours each day. Like all forms of programming, the range of light entertainment was restricted, as was the general audience for television. In July 1952 the television public constituted 14% of the population as a whole (Silvey, 1974, p.164), which amounts to approximately one set in every eight homes. Just after *Current Release* began, the average evening audience was just over two million, while nearly ten million listened to radio.[22] It is widely perceived that television was available to a more mass audience in the period 1954 to '56 when ITV was introduced, although the BBC's coverage of the Coronation in 1953 (watched by an estimated 20 million people as compared to the 12 million who tuned in to radio), is also considered to be a watershed in this respect (Stokes, 1999, p.33). However, at the beginning of the 1950s a 'budget set' cost £50 against an average weekly wage of £7 (Corner, 1991, p.2), and in 1952 hire purchase was temporarily suspended (Briggs, 1979, p.245). A reader in *Picturegoer* was perhaps expressing the views of many when she complained that 'Television is too dear for the average picturegoer'.[23] *Current Release* emerged at a time when television was just beginning to expand but (unlike *Picture Parade* and *Film Fanfare* later in the decade), it was not necessarily accessible to a mass audience.

Prior to this, 1947 saw the advent of *Film History*, 'A series based on the early days of the cinema going back to the old silents',[24] and *The Film* (1947-8), a profile centred programme with editions on esteemed directors such as D. W. Griffith and Sergei Eisenstein. *The Film* was introduced by Roger Manvell and produced by Andrew Miller-Jones, both of whom were established film critics, and the BBC also screened isolated programmes such as *Art and the Film Camera* (1950) and *The Birth of the Cinema* (1950). These examples - with an emphasis on film as an 'art' - are primarily shaped by a bid to educate the audience, or at least 'inform' them of aspects of film culture unfamiliar to them, either by way of historical distance or national context. While indicative of the BBC's wider interest in exploring the contexts and history of film culture, they could hardly be seen as precursors to the more popular and contemporary *Current Release*. From 1946 to '52, film stars often appeared on *Picture Page,* but clips from the films were seldom made available.

With regard to the sale of feature films, the film industry's initial hostility towards television feature films was understandable for, while it was not unusual for the British film industry to be plagued by financial problems, it began the decade in less than robust shape. Cinemagoing experienced a boom period during wartime and critics had also celebrated an artistic 'renaissance' in British film. During the war Sir J. Arthur Rank consolidated his hold over the British film industry and financed many successful productions, but with his ambitious bid to conquer the American market, he had severely overstretched himself. As the Entertainment Tax continued to sap the revenue of British cinema, Rank announced a huge overdraft in 1949 which some perceived 'brought the entire British industry to its knees' (Macnab, 1993, p.193). Rank successfully retrenched his company and restructured his assets, centralising limited production at Pinewood, but critics offer differing accounts of the industry's wider economic health in the 1950s. Robert Murphy stresses that the British film industry 'enjoyed a greater health and stability than might have been expected given the competition from television' (1989, p.230), and as he points out, it produced a number of successful mainstream films at this time. Clearly, however, as Christine Geraghty points out, this 'relative stability' needs to be situated in the context of the changing social significance of the cinema on a wider scale - not least of all the downward spiral in admissions (2000a, p.20). Other critics certainly paint a less robust picture of the industry (see Macnab, 1993, 2000), and *Picturegoer* began the decade by noting with disapproval that 'The squandermania of the past has caught up with the [British] studios in a big way. We, the public want to see their films, but they haven't the cash to make them'.[25] The Government did institute measures to strengthen the ability of producers to secure a reasonable return at the box office, but the British industry was again plagued by limited cash flow. With respect to the exhibition sector, the difficulties posed by the general decline in admissions were not helped by a shortage of product from Hollywood when, following the reorganisation of the studio system, the majors concentrated on fewer, more costly pictures (see Balio, 1990).

As discussed in Chapter One, by 1952 cinema admissions had already begun to decline. While television was not the sole contributing factor here, the film industry were very aware of its competition, and their initial dealings with the BBC were characterised by an attempt to control the new medium. In the late 1940s and early 1950s there were discussions between the BBC and the Rank Organisation on the subject of large-screen or 'Cinema-TV' in which television would be delivered via a large-screen format to theatrical audiences (see Buscombe 1991). While *Current Release* was on air the trade papers were littered with news on the development of Cinema-TV and in 1953, Rank and ABPC filmed and transmitted films of the Coronation for theatrical exhibition. As Chapter One outlined, as far as the film industry was concerned, the institutional and technological future of television was still taking shape, and they had no desire to further cement its growing centrality as screen entertainment in the domestic sphere.

Just before *Current Release* began, the BBC's Controller of Programmes, Cecil McGivern, acknowledged that, 'There is practically no chance of our obtaining good full-length films either made or distributed in this country for a long time to come'.[26] This was thus a time when it became increasingly important to foster workable relations with the film industry, particularly as television hours and audience numbers gradually increased. McGivern emphasised just before *Current Release* began that 'This is a most important series and our future relations with the film industry will depend ... on [its] success'[27] - again acknowledging the cinema programme's status as a form of institutional dialogue between the media. Formal discussions on the project began in June 1950, and central here are the processes, the decisions and indecisions, which came to define *Current Release* and the various aesthetic, institutional and cultural factors which shaped its formation. This is particularly so given that the early discussions over the programme set institutional and textual parameters within which the genre would function for many years. *Current Release* was shaped by the history of the BBC's institutional relations with film culture while, as a television programme, it simultaneously brought new possibilities, conventions and problems to the fore.

The negotiations over the programme were primarily handled by the Better Business Committee which was formed by joint representatives from the CEA (exhibitors) and the KRS (distributors), collectively also referred to as the Film Industry Publicity Circle (FIPC). Although the description was somewhat vague, *Kine Weekly* explained how the FIPC was formed to handle 'public relations and general industry publicity'.[28] With regard to the BBC, the most important players in the drama were Philip Dorte head of the BBC Film Department, W. Farquarson-Small, the producer of *Current Release*, and Cecil McGivern, the controller of Television Programmes. As Briggs describes, the terms of McGivern's appointment were 'to supervise the overall quality of television output' (1979, p.224) through which he 'retained a large measure of personal control over programme detail' (Black, 1972, p.145). At the level of a specific text or case study, this supervisory capacity is certainly made clear by the development of *Current Release*.

In planning the series there were two issues which dominated the discussion, and they suggest the key debates which surrounded the cinema programme from its inception. The film industry were largely concerned with the verbal rhetoric of *Current Release,* the extent to which it would have scope to *criticise* the films, while the BBC's prime concern was the problem of 'advertising', and thus the degree to which *Current Release* could be seen to *publicise* the cinematic product. As already established, these conflicts were raised in the pre-war negotiations and in fact played a key role in preventing a cinema programme from going on air. When it came to *Current Release*, these issues were rapidly seen to operate in dialogue. From the BBC's point of view, if the programme had a critical angle, it would reduce the impression that it was advertising the films, yet the film industry were only too aware of this and desired that their products be publicised to the fullest extent. From this point on, what were perceived as irreconcilable concerns were firmly opposed: freedom of criticism had a long history in relation to public service, while 'advertising' was technically a violation of the BBC's official charter. This was clearly articulated by McGivern when negotiations with the film industry began when he explained how the Corporation hoped to produce something 'which would be of interest and enlightenment to viewers... It would not be possible to entirely avoid the introduction of criticism' because otherwise, 'the public would criticise the BBC, alleging that it had made a deal with the industry to help exploit its films'.[29] At this stage it seemed that the investments and aims of the two parties were not only opposed, but apparently irreconcilable.

Television has always faced considerable restrictions where the subject of film criticism is concerned, so these discussions mark a crucial point in the development of the genre. In discussing his experience of producing *Moving Pictures* in the years 1990 to '93, Paul Kerr argues that:

> TV ought to be the ideal medium for discussing cinema for the simple reason that one can 'quote' rather than, as in the case with the print medium, being obliged to summarise the scenes or sequences being discussed... But... in making the transition from page to screen, 'film criticism' on television is subject to considerable constraints which simply don't apply in print

(1996, p.140).

In discussing *Moving Pictures*, Kerr explains how access to film excerpts, interviews and other information is subject to many financial and legal restrictions and he describes the 'unwritten code of PR etiquette' which prohibits 'disrespectful' reviews or interviews (1996, p.142).[30] Although the press and radio can equally come under attack, television has faced the most severe restrictions in this respect. Perhaps this is precisely because of its similarity to the cinema - what Kerr perceives as its ability to 'quote' directly from the film text - and hence its apparent publicity value. As the *Daily Film Renter* noted in 1952, *Current Release* was even better 'publicity ... than the multitude of radio programs [sic] which... place the accent on films. Visual publicity will always be better than aural as far as the screen is concerned'.[31] Exploring the emergence of the genre, however, enables a consideration of how and why these conventions were established and the institutional, aesthetic and ethical considerations which surrounded them. From the very start, the cinema programme made explicit certain tensions at the *core* of broadcasting policy and the BBC's conception of their wider rhetorical function. And it is not difficult to see why. In terms of television in general, the BBC stated that programmes should 'fulfill functions corresponding to those of an expert critic in a newspaper, as distinct from the functions of a newspaper's advertising columns'.[32] The dialogue between these discourses was quite *literally* situated at the very centre of the cinema programme. Much of the BBC's policy on 'advertising' had been formed with regard to radio, but the visual nature of television was perceived to increase the difficulties this posed.[33] It is further important here that films were deemed to be an exception to the rule, an area of latitude where 'common sense' should prevail. Trade names, for example, were generally subject to exclusion from radio and television programmes, but the mention of film companies was permitted. This was because they were considered to be of 'general usefulness' to the viewer and the product of the film companies was part of the programme itself. Similarly, as on radio, information could be given about general or local releases of films, and celebrities were permitted to discuss their present plans 'quite freely', although they were nevertheless to be made aware of the dangers of 'over-plugging' their product before going on air.[34] *Current Release* was to operate within an area in which a certain amount of promotion was inevitable, although carefully policed. The main (and rather vague) guideline was that such 'advertising must be limited by the consideration of what is of *general* interest to viewers'.

It is important to stress here, however, that *Current Release*, and the cinema programme throughout the decade, in part occupied a heated site of controversy where 'advertising' was concerned because of the wider debates circulating around the relationship between television and promotion at this time. While the BBC had always been aware of, and vehemently resisted, the alternative of advertising as a means of funding broadcasting (see Briggs, 1965, 1979), the advent of commercial television in Britain made this a reality. Although ITV did not emerge until 1955, debates over the introduction of a commercial television service began after the war, effectively permeating BBC consciousness since the late 1940s. Graham Murdock emphasises how 'there was a widespread feeling within the establishment that a television service funded by advertising was thoroughly alien and unBritish and would undermine public broadcasting's mission to build a common national culture' (1994, p.204), and the form that a commercial television might take in Britain was the subject of anxious debate in the years leading up to 1955. With the spectre of Americanisation looming large, there had been a great deal of public and governmental concern that 'sponsored' broadcasting would become a reality in Britain, right up until the Television Act of 1954. Even after the Television Act announced the decision to implement spot advertising in 'natural breaks', there remained a great deal of public confusion over the matter (Sendall, 1982, p.98). The first edition of the *TV Times* was at pains to distinguish ITV from American commercial broadcasting, assuring the reader that the advertisements would be unobtrusive and in fact enhance viewing pleasure, combining 'information and entertainment'.[35] In particular, there was concern that a clear boundary between advertising and programme material be maintained (Sendall, 1982, p.98). This was articulated in opposition to the American broadcasting system given that, up until the late 1950s (when 'spot' advertising was also introduced), programmes were produced by advertising agencies, with the networks themselves having little creative input (Hilmes, 1990, p.82). This created an economic and textual context for such strategies as the 'integrated commercial message': 'an advertising plug arising so smoothly out of the program action, or actually written into the narrative, that it was indistinguishable from the dramatic structure' (Hilmes, 1990, p.86). For example, this had particular implications for the coverage of film. *Lux Radio Theater of the Air* promoted Hollywood stars and films while endorsing Lux soap - by both direct and indirect means. While references to the product were integrated into scenes and sketches, host Cecil B. DeMille would also explain how 'Lux Toilet Soap is the official soap over on the Paramount lot and every other great studio in Hollywood' (Hilmes, p.94). Such intertextual promotion was of course not permissible in the British context, whether in relation to film culture, or the wider contexts of broadcast material. Emerging with the advent of ITV, the genre of the advertising magazine, for example, clearly blurred the boundaries between advertising and programme material.[36] Yet after causing considerable debate, the form was ultimately outlawed by the Pilkington Committee in 1962 when it was decreed that it broke the terms of the Television Act which stipulated that advertising be extraneous to programme content (Sendall, 1982, p.213).

This wider context in the 1950s goes some way to elucidating both the institutional concerns over and reception of the cinema programme at this time. The recurrent criticism of the genre as a reprehensible form of 'plug TV' was significantly also a popular phrase associated with the concerns surrounding ITV. Much of the concern around the introduction of advertising to British television pivoted on concerns over the perceived 'suggestibility' or 'gullibility' of the television audience (Thumim, 2002b, p.211). The institutional concerns which circulated around the cinema programme, however, seem to be organised more around the cultural value of television and in particular, the 'moral' reputation of the BBC (recall the worry that 'the public would criticise the BBC, alleging that it had made a deal with the industry to help exploit its films').[37] A perennial problem for the BBC throughout the 1950s was that the cinema programme necessarily blurred the boundary between promotion and programme material, functioning in what can be seen as an ambiguous institutional and textual space.

These debates were played out from an early stage. A year before *Current Release* began the FIPC submitted a pilot script to the BBC called *Filmtime*. Given that they were not going to be directly producing the series, this was presumably to illustrate how they thought a cinema programme *might* be conceptualised. The BBC were not at all impressed by the script and one of its main problems was indeed perceived to be excessive promotion.[38] McGivern insisted that there was 'far too much direct advertising... almost sponsored advertising... it advertises companies, producers, and the cinemas which are showing the films'.[39] The BBC considered the script to present precisely what they most feared, a programme of 'trailers'. The film industry were equally vigilant where the issue of criticism was concerned. Despite McGivern's reassurances that such verbal comment 'need not be destructive or unkindly' and that it was not the BBC's intention to 'harm the film industry',[40] they remained unconvinced. According to the KRS, they believed it would 'be better ... for the subject to be dealt with "factually" from the point of view of television'[41], an approach which presumably favoured a descriptive, as opposed to evaluative rhetoric, and they mentioned what they had considered to be previous problems concerning film criticism on radio. That is not to suggest that the film industry's relations with radio had been largely antagonistic. They had clearly valued the publicity that radio film reviews (and other features) could offer, and the extent of this coverage was considerable. Nevertheless, particularly around the film reviews, there were several occasions when tensions arose. In 1945, for example, Gaumont British undertook official proceedings when a BBC critic was accused of 'defaming' one of its films. The company angrily informed the BBC that the broadcast:

> went well beyond fair criticism... and amounts to an attack on the British film indus-
> try. While Rank is in the USA trying to sell British films, the BBC's critic is saying
> that she doesn't know why [the] films were made.[42]

What is particularly significant and telling here is that the BBC considered this to show a 'complete misunderstanding of the purpose of our film reviews'.[43] This crucially indicates how the BBC and the film industry had very different perspectives on what 'film criticism' actually *was* (and as discussed below, the Corporation often drew upon the discourses of professional British film criticism). From the BBC's point of view, this purpose had two important links with the policy of public service. The reviews were first and foremost intended to help the listener to select the pictures they wished to see, but they were also intended to raise their critical appreciation of film. As the CEA were informed after objecting to a review:

> We leave our critics as much freedom of criticism as possible and we find that this
> results in broadcasts which help the listeners in their cinemagoing, which make
> good programmes, and which are a contribution to the formation of standards
> among film[goers].[44]

This clearly suggests that freedom of criticism was seen as integral to the service function of the review, as well as its educational and entertainment functions. In this respect, then, film criticism had already 'raise[d] a number of points of BBC policy',[45] but it rapidly emerged that television was to exacerbate such tensions, and bring further problems to the fore.

None of the cinema programmes in the 1950s could be seen as offering a critical review of the films on offer. *Current Release* ultimately included little negative criticism of the films it screened and the correspondence indicates that the BBC envisaged a more critical approach than actually transpired. Yet in considering the shift of film culture to television, what is significant is that this critical approach appeared to be perceived as *less* integral to a televised cinema programme. From an early stage, McGivern suggested that they might 'avoid the established newspaper and periodical critics' and offer 'an element of commentary by representatives of the ordinary filmgoer'[46] - a decision which raises a number of issues central in understanding the institutional, aesthetic, and textual factors which shaped the development of *Current Release*. What does the decision to use an 'ordinary filmgoer' suggest about the relations between the BBC's conception of its audience and the principles of public service? What does it indicate about the BBC's relations with popular film culture? How might this decision be shaped by the specific demands of television, and its developing modes of address?

Film on Radio: To Educate and Entertain

Firstly, McGivern's suggestion that the presenter should be a 'representative of the ordinary filmgoer' has a democratic inflection, and suggests an aim to address a mass public. The concept of democracy is perceived to have a complex and shifting relation with public service (see Crisell, 1997, pp.27-30). Beginning with radio, the BBC's policy of mixed programming was intended to offer something for everyone, yet it is widely known that the Corporation was accused of being undemocratic and elitist. Andrew Crisell argues that the question of elitism arose because the BBC's public service policy strove to provide the 'best' of everything (music, drama and so forth), yet this is implicitly couched within class prejudice, and notoriously difficult to define (1997, p.28). The BBC's tastes may not have concurred with those of its audience, although Reith's perception of the relation between broadcaster and audience had little sympathy with the view that this match was desirable. Clearly, his suggestion that the BBC should lead rather than 'pander' to public taste is widely known.

Radio's coverage of film fell into three broad categories of reviews, talks and programmes, the scope, content and address of which differed in the pre and postwar periods. It was with the film talk (which grew out of the review in the 1930s), that there was perhaps the clearest bid to 'educate' the listener. The film talk covered a wide range of topics from the aesthetic, artistic to the industrial. When they first began their intention was to 'bring listeners to a proper conception of the principles of the art concerned',[47] and the BBC suggested that 'the most hopeful line of approach the teaching of the public how to look at films, to recognise the special quality and habits of various producers'.[48] Rather than an early intervention in the 'auteur' debate, this is a conventional conception of education in the arts: art is made by artists, and it is here that audience appreciation should begin. It can be seen here that, although the BBC's relations with film culture began at a time when the cinema was the central mass entertainment medium, the growth of the commercial cinema had for some time existed alongside the claim that film was 'art', and thus that it deserved to be taken seriously as an 'art' form. As Tom Ryall explains, in addition to the increasing attention the medium received from the serious press, 'a serious critical and theoretical literature of cinema began to emerge' (Ryall, 1996, p.13). British intellectuals of the 1920s and 1930s were critical of commercial filmmaking and tended to favour avant-garde and European cinema, or the British documentary movement (Richards, 2000, p.24). These interests, of course, were representative of a specialised and minority film culture, and were not part of British cinemagoing for the mass audience. Yet they are important in understanding the discourses on which the BBC may have drawn when the cinema was regarded as primarily a mass entertainment form. The BBC clearly understood the need to conceive of the cinema as predominantly a popular entertainment medium, yet they seemed to offer a range of material which was situated between the film fan magazine (interviews with the stars, coverage of current films) and the discussion of film in the quality press. As discussed below, they frequently used critics from this sphere.

However, that said, it is difficult to describe the BBC as having a particular 'approach' to film given that this was relative to particular programmes, forms and different historical periods. For example, from the mid- to late 1930s, critics have described a softening in the BBC's didactic approach and 'an increased sensitivity to public taste' (Crisell, 1997, p.30). This was due to a variety of factors (such as competition from commercial stations, Reith's departure and the emergence of BBC Audience Research), and it gathered ground during the war. As Cardiff and Scannell explain, the restructuring of radio during the Second World War made the medium 'more attractive and accessible to the working class, and more in line with their tastes and requirements' (1981, p.41). They argue that the wartime period initiated a decisive shift in BBC broadcasting, although it was also an acceleration of trends apparent in the late 1930s. This process was not only a move toward more popular radio, as it expressed what Cardiff and Scannell describe as a 'redefinition of [the BBC's] ... relationship [with] its audience, or rather its acceptance of the audience with different tastes and needs' (1981, p.75). It was the recognition of a more diverse audience which gave rise to the post-war emergence of the Home, Light and Third programmes on BBC radio. The Director General, Sir William Haley, conceived of these programmes as a 'broadly based pyramid, slowing aspiring upwards' the intention of which was 'to lead the listener up the cultural scale' (Sinfield, 1989, p.51) (the Third programme representing the 'highbrow' approach). This move, controversial within the BBC, can be interpreted in different ways. It can be perceived as the BBC's most vehement and organised attempt to 'improve' the nation's taste or, alternatively, as in part a rejection of the mixed programming policy which had previously been its aim. Either way, these shifts impacted upon the relations between radio and film culture and, subsequently, the existing institutional context from which *Current Release* emerged.

In the post-war period with the development of the Home, Light and Third programmes there emerged a reorganization of the BBC's coverage of film culture. We see what might be termed a 'stratification' of the coverage across the three programmes but, also, with the advent of the Light programme, an increasing *popularisation* of approach. Film talks appeared on all three services, and even addressed themselves to the minority audience of the highbrow Third. To take the year 1952 when *Current Release* emerged, the most important example on the Home service was the long-running series *The Critics*, which also included talks on theatre and literature. Although often dealing with current films, it was not simply a review but was intended as 'a critical discussion of the most important film available during the week' in which the critic was described as 'either a professional... or someone well-acquainted with the artistic objects and technique of filmmaking'.[49] In addition to features in *Woman's Hour*, the Light programme included such examples as *Younger Generation* which featured discussions between 'under-twenties' and writers or film directors. The purpose was to describe 'the objective of the filmmaker and how he fulfils it' and the conversations were described as 'critical and outspoken'.[50] The Light programme also broadcast *Under Twenty Parade* which, evidently aimed at the youth audience, offered 'comments by young people on visits to [the film] studios', including occasional interviews with directors. In contrast, the coverage of film in the Third programme was a subject of controversy within the BBC, as there was a question as to whether (even in its more 'serious' forms) it was a worthy enough subject for such a cultured sphere. While the Controller of Talks would repeatedly insist that the Corporation had 'significantly underestimated the amount of serious interest in the cinema today',[51] he was informed that 'the major interest is in the popular film and the cinema star' and that the 'number of those seriously interested in "the film"' was likely small, even in the minority audience for the Third programme[52] (which attracted approximately 1% of listeners in comparison to 40% for the Home service, and 60% for the Light) (Cardiff and Scannell, 1981, p.69). Nevertheless, the Third programme broadcast a number of film talks on such topics as 'The cinema and the unconscious', 'The British Quota system', 'The Indian film industry', 'Shakespeare and the cinema', or discussion on the theories of Russian filmmaker Sergei Eisenstein.[53] But the BBC's emphasis on the extent to which the 'popular interest' in film should predominate was in itself indicative of a certain shift. Essentially, this appeared to place the focus more squarely on entertainment, and the advent of film magazines in the late 1940s epitomised this shift.

The film programme was broadcast on both the Home and Light services (although predominantly the Light), and it covered a wide variety of features. These included series such as *Movie Matinee* and *A Seat in the Circle* which, largely focusing on current pictures, presented excerpts or adaptations from film soundtracks. In the late 1940s and 1950s, there were also magazine programmes such as *Filmtime* (Home), *Movie-go-Round* (Light) and *Picture Parade* (Light). *Filmtime* included such items as 'Can you identify this film?', 'Picture Panel' in which stars and producers answered listeners' questions, and 'The part I would most like to play' in which a film star enacted a scene of their choice. It also included visits to stars' homes, trips round the film studios, and general excerpts from current films. *Movie-go-Round* emerged in 1956 and the forty-five minute feature began with music from a new release, then moving onto 'Picturegoer's

Quiz' and 'Around the British Studios'. It would then hand over to its 'Hollywood Correspondent', Donovan Pedelty (who performed a similar function in *Picturegoer* at this time). This was followed by items such as 'Soundtrack Memories' and adaptations of particular scenes. As this indicates, there was understandably an attempt by radio to exploit the *aural* dimensions of film (and in this respect it seems significant that radio's coverage of film culture initially developed at the time that the cinema 'found its voice' in the late 1920s) (Hilmes, 1990, p.53). Beginning in the late 1940s, radio's *Picture Parade*[54] included similar features to both *Filmtime* and *Movie-go-Round*, although it also offered the intriguing features 'Picture Post Box', 'The best suggestions of the week for the improvement of British pictures' and 'How not to make a movie', which took a satirical look at a specific theme or topic.[55] In the first edition this focused on the 'lack of authenticity in costume pictures' by courtesy of 'Drearitone Pictures Inc.', while the second edition discussed examples of ineffective and misleading film publicity. This outline makes clear that these programmes were primarily based on a desire to entertain, but they nevertheless encouraged a lively discussion on all aspects of film culture - often with a critical edge. As *Kine Weekly* ruefully recalled in 1952, 'Many people in the trade have bitter memories of that cynical BBC [radio] programme *Picture Parade*. It ridiculed the film industry'.[56]

These programmes may well be evidence of the 'softening' in the BBC's approach and their 'increased sensitivity toward public taste' (Cardiff and Scannell, 1981, p.41), and they suggest a self-conscious attempt to address a wider audience. Compare, for example, the outline of the film programmes above with the film industry's complaint in 1935 that the BBC's film criticism 'was above the heads of all except a negligible minority' with a 'tendency to talk to Chelsea and Bloomsbury and to praise the arty Continental films'.[57] In comparison, the producer of radio's *Picture Parade* explained that he hoped to cater for 'the film fans, the normal picturegoer who goes because he [sic] wants to see a certain film.... and the more technically-minded picturegoer who goes to study and criticise a picture'.[58] This was in fact suggested by the textual address of the programme which was shaped by these 'imagined' constructions of the listening public. One of the speakers, for example, was a 'Miss Film Fan' who explained how: 'I'm what so many of you are, a film fan. I work hard all day in my office. I go to the flicks to be entertained - and I most certainly know what I like' - suggesting a very self-conscious attempt to meet the envisaged 'needs' of the audience. This also indicates that while the creation of the Home, Light and Third programmes enforced a certain stratification of the listening public (from the BBC's point of view), individual programmes aimed for a *broad* appeal. After all, an integral aspect of public service as defined by the BBC had always been 'the range of interests catered for' (Crisell, 1997, p.46). At least at the level of film culture, this did not appear to diminish with the development of the Home, Light and Third programmes, despite evidence of a more stratified approach.

Certainly, the cinema programme on television in the 1950s was largely focused on popular film culture and previews of the latest films, as the title of *Current Release* suggests. While it retained a desire to educate and inform as well as entertain, *Current Release* emerged at a time when there had been a popularisation of the BBC's coverage of film culture, and the recognition of a more diverse audience. McGivern's suggestion that it should be presented by (and presumably address) a 'representative of the ordinary filmgoer', emerges at a time when the BBC's conception of this entity had shifted somewhat. Similar to the producer's description of radio's *Picture Parade* (although more elitist in tone), McGivern suggested that *Current Release* should appeal to 'the intelligent filmgoer, as well as the bobby-soxer'.[59] Although radio's film programmes had also aimed for a broad appeal, the televised cinema programme did not offer the opportunity to address 'separate' audiences (as conceptualised by the BBC). This need to attract an undifferentiated mass audience was arguably more pronounced with television, and may partly explain why, in comparison with radio, its coverage of film has historically been more popular in tone. That said, we have seen how television in 1952 was not available to such a wide public, although it is possible that it was at this point that a proportion of working-class families began to acquire a set. (Following a survey in 1952, BBC Viewer Research concluded that while 'the possession of a TV set is still highly correlated with income... well over half the TV sets in use last May were to be found in the homes of the working class'.)[60] This may also have shaped McGivern's decision to use 'a representative of the ordinary filmgoer' to present *Current Release*, as well as influencing the decision to produce a televised cinema programme at all at this time.

Yet it may be misleading to become too preoccupied with issues of class in this respect. While they may be a structuring force in the BBC's own conception of its audience, such categories are rather homogeneous, intangible and reductive. A class analysis may also obscure how the decision of who should present *Current Release* was perhaps shaped by more aesthetic concerns, particularly the perceived demands of a televisual mode of address. Kerr notes that unlike television's treatment of other arts, 'Cinema programmes haven't been deemed to demand specialist presenters' and relates this to the status of film as a 'popular art' (1996, p.139). However, in situating this within the development of the genre, certain questions are raised. Radio's interaction with film culture was most intense in the 1930s and 1940s, the heyday of cinema as a popular medium, yet it often used specialist presenters. In the late 1940s the Corporation made regular use of *professional* film critics such as Dilys Powell, Arthur Vesselo, Basil Wright, C. A. Lejeune, Roger Manvell and Ernest Betts. Several of these wrote for periodicals such as *Sight and Sound* and *The Penguin Film Review* and regularly wrote for the national press. Canonised as a pantheon of 'middlebrow film journalism' in Britain (Chapman and Geraghty, 2001, p.3), in the 1940s they were what John Ellis (1996) has subsequently described as 'quality' critics, in that they participated in the creation of a critical discourse by which British cinema of the period was judged (primarily the values of documentary realism, a middle-class improvement ethic and literary quality) (Richards, 2000, p.24).

It is certain that the perceived needs of the 'ordinary filmgoer' were considered in relation to radio, but this was apparently deemed to be more important where television was concerned. In discussing a presenter for *Current Release*, McGivern not only hoped to convey 'an element of comment' by the 'ordinary filmgoer'. He also envisaged a speaker 'whom the public would wish to hear and see'. John Corner has referred to television's developing modes of address as operating through a 'rhetoric of *understatement,* of the *self-evident* and the *implicitly shared* "Usness"' [original emphasis] (1991a, p.12), which drew upon, yet extended, the conventions and possibilities of radio. It is clear from the discussions surrounding the *Current Release* that there was an attempt to create a relaxed informality suited to the specificity of a visual medium. The decision to use a non-professional film critic was also partly shaped by the 'no criticism' clause (as time progressed the producer acknowledged that an established critic would be inappropriate as 'most worthwhile [critics]... would regard [it] as a dangerous devaluation of their professional status').[61] It was also shaped by the fact that the BBC were keen to build their own critics. Nevertheless, it would seem that the emerging conventions of a 'televisual' mode of address also influenced this decision. As based on a rhetoric of familiarity, casualness and informality, this is crucial in considering how the wider discursive function of the genre has developed.

The presenter ultimately selected was thirty-three-year-old John Fitzgerald, a name which was to become increasingly associated with cinema programme in the 1950s. Although by 1952 he had played very minor parts in film and television, Fitzgerald was likely not a familiar face to viewers. The producer felt that he was appropriate in the context of the desire to offer:

> intelligent presentation rules which out any form of clown or mere jester, however gifted or popular... But 'without criticism' rules out the obviously superior, critical personality. To introduce a Dimbleby at this stage, with present limitations, would be to put him and us and the viewers - all in a false and embarrassing position.[62]

John Fitzgerald was evidently perceived to lie somewhere between the 'clown' and the 'superior' personality, and although McGivern continued to express concern that he was too 'lightweight' to deliver the intelligent approach required, it was his occupation of this middle ground which made him so appropriate.

When they spoke of finding the 'right man for the job' the use of the male adjective was entirely appropriate, since at no point was the idea of a female presenter entertained. Based on the evidence of the memos it would appear that there was no discussion of what Janet Thumim has elsewhere called a 'self-evident policy decision' (1998, p.92). Joy Leman explains how within the BBC, a recurrent debate had focused on women's 'inherent unsuitability to the medium of radio and later to major areas of television' (1987, p.79). As Head of Television Current Affairs in the 1960s, Grace Wyndham Goldie was one of few women to obtain a position of power in the BBC in these early years, and she had her own theory as to why female presenters were a minority: 'Women's diction... produces a more class-conscious connotation. Listening to a woman on TV, one can almost put her into one class or another, but this doesn't apply to men...' (cited in Leman, 1987, p.79). Despite perceptions of the BBC's middle-class tone, she implies that the Corporation sought a 'classless' rhetoric which women were less able to achieve. This desire for a broad address is evident with respect to *Current Release* (the presenter as 'ordinary filmgoer'), but the preconceptions which structured women's marginalisation in early British television seem to be more complex than this.

The decision to use a male presenter was arguably shaped by the more general conception that a woman would not carry the necessary authority and 'weight'. Although some of the best remembered announcers in early television - with the privilege of speaking direct to camera - were women (such as Sylvia Peters and Mary Malcolm), female TV presenters in the 1950s were usually associated with a limited range of genres such as magazine, 'women's' or children's programmes. In these early days of television women were not, for example, perceived to have sufficient gravitas to read the news, and when the BBC experimented with this possibility (as late as 1960), the thought that a woman 'could be the conveyor of truth and authority' was met by resistance in the Corporation (Thumim, 1998, p.97). *Current Release* was primarily an entertainment programme, but the role of the presenter was nevertheless regarded as one of casual *authority*, and this command was seen as all the more important given the tensions surrounding the 'no criticism' clause. In subsequent years the BBC's *Picture Parade* occasionally included a female interviewer (Elizabeth Havelock), and the later editions of *Film Fanfare* were co-presented by Macdonald Hobley and Patricia Lewis. Yet even on these occasions, women were evidently selected as much for their physical appearance as professional ability (their looks were constantly commented on by the male writers in the film trade press), and in subsequent years the presentation of cinema programmes has remained an area dominated by men.

This decision may seem contradictory as two of Britain's leading film critics at this time who broadcast regularly on BBC radio were women - namely Dilys Powell and Caroline Lejeune. This discrepancy is crucial as it suggests that it was less about the presenter commanding an *expert* knowledge of film (which none of the male presenters in the first ten years of the TV cinema programme possessed), than it was perhaps also shaped by the BBC's perceptions of television's developing *visual* modes of address and their relations with authority, control and informality.

Particularly when using a non-professional critic, the role of 'a knowledgeable and personable' presenter was evidently perceived as best entrusted to a male. Equally, as discussed further in the analysis of *Current Release*, there is evidence to suggest that, when it came to popular film culture, women were more likely to be perceived as associated with the discourses of film *fandom* - signifying a 'frivolity' and immersion within film culture that the BBC were so keen to avoid. While the decision to use a representative of 'ordinary filmgoer', then, evokes connotations of democracy and consensus (and it *was* shaped by a well-meaning desire to meet the interests of audience), this also has its negative counterpart. The gendered address of the programme is considered in more detail in Chapter Three, but from the BBC's point of view, a white, middle-class, and above all, *male* viewpoint tends to colonise the definition of the normal and the 'ordinary'.

This chapter has established the discursive context from which *Current Release* emerged, a context which, in terms of the dialogue between the film industry and the BBC, was characterised by a struggle over their differing investments in the programme, and conflicting expectations about how it should develop. This dialogue ultimately defined a complex space within which the early cinema programme would function - central aspects of which would remain in place not only for the decade, but for much of its history. Following eighteen months of negotiation (and the production of a successful pilot edition by the BBC), the final formula for the series was set. Less a programme of criticism than the 'intelligent presentation of selected excerpts', *Current Release* would be a forty-five-minute feature screened on a fortnightly basis. It would deal with films on general release, and the BBC would retain complete control over its production - including the selection of the films and excerpts. *Current Release,* 'the first official link between the BBC and [the] local cinema',[63] was ready to go on air.

Notes

1. *Daily Express*, 17 March, 1973. BFI press clippings folder on cinema programmes.

2. *Tribune*, 20 November, 1978. BFI press clippings folder on cinema programmes.

3. *Daily Mirror*, 5 June, 1952. BBC press cuttings, file 3A.

4. *Daily Star*, 9 January, 1953. BBC press cuttings file 3A.

5. Daily Express, 29 December, 1952. File 3A.

6. 'Film': undated memo prepared by Gerald Cock. WAC T16/72/2.

7. 11 June 1935, letter to BIP from Gerald Cock. T16/72/1.

8. 2 March 1936, letter to RKO from Gerald Cock T/240/1.

9. 10/4/46 Memo by Director of talks. R51/173/4.

10. As above.

11. 2 March 1936. Gerald Cock to RKO. T/240/1.

12. As above.

13. 11 June 1935. Gerald Cock to British International Pictures. T16/72/1.

14. As above.

15. 13 October 1937. 'Feature Film', memo by Gerald Cock. T6/138.

16. Although given that the backbone of the cinema audience was working class, and that the audience for 1930's television was extremely limited and likely upper class, it is perhaps questionable as to how popular such a programme would have been.

17. 19 December 1936. T6/194.

18. 7 March 1939. Myers to Gerald Cock . T6/194.

19. As above.

20. 17 December 1937 Cock to KRS. T6/138.

21. 8 March 1939 Cock to Myers. T6/194.

22. The Times, 12 March, 1952, p.10.

23. Picturegoer, 1 March, 1952, p.8.

24. Radio Times, 11 May, 1947, p.49.

25. Picturegoer, 20 January, 1951, p.5.

26. 13 December 1951 Cecil McGivern to Director of Television. T16/72/4.

27. 28 November 1951. Cecil McGivern to Director of Television. T6/104/1.

28. Kine Weekly, 14 February, 1952, p.4.

29. 1 June 1950. Meeting between BBC and CEA/KRS. T6/104/1.

30. In more recent years the film industry has taken increasing editorial control over this area and the Hollywood majors freely distribute Electronic Press Kits (EPK's) which are pre-prepared celebrity sound bites with approved clips and trailer (Kerr, p.141). Unless the television company is willing to pay considerable costs, this is the material they have to use, whether it is suitable for their particular programme or not.

31. Daily Film Renter, 15 May, 1952, p.4.

32. Undated 'Draft policy on advertising in programmes'. R34/1.

33. Undated memo on 'The advertisement and the sponsoring problem in television programmes'. R34/1.

34. May-July 1954. 'A paper describing present practice with regard to "advertising" in BBC programmes', prepared by Cecil McGivern. R34/963/1.

35. TV Times, 20 September, 1955, p.3.

36. In the ad-mag, several short but related items were linked by a presenter/ host. Thumim, for example, refers to the example of Jim's Inn (AR, 1957-1963) which had a couple running a village pub in which they 'discussed the prices and quality of various domestic products with their customers' (Thumim, 2002b, p.211).

37. 1 June 1950. Meeting between BBC and CEA/KRS. T6/104/1.

38. The script, entitled Filmtime, is held in the Current Release files. T6/104/1.

39. 19 March 1951. McGivern to John Dennett. T6/104/1.

40. 1 June 1950. Meeting between BBC and CEA/KRS T6/104/1.

41. 25 August 1951. KRS to McGivern. T6/104/1.

42. 6 July 1945. Gaumont British to BBC. R51/173/4.

43. 13 July 1945. Controller of Home service to Director of Talks. R51/173/4.

44. 26 November 1941. BBC to CEA. R51/173/3.

45. As above.

46. 1 June 1950. Meeting between BBC and CEA/KRS. T6/104/1.

47. 20 February 1933. Memo on film. R51/173/1.

48. 23 October, 1936. Memo on film. R51/173/1.

49. 6 February, 1952. Chief Assistant Talks to CT. R51/173/5.

50. As above.

51. 31 August, 1949. Michael Bell to C.T.R51/173/5.

52. 6 July, 1949. Harman Grisewood to CT. R51/173/4.

53. Undated proposal for 'Film talks in the Third programme'. R51/173/4.

54. This is not to be confused with the television programme of the same name that occupies a central focus of this book. It appears that the BBC simply adopted the name of a film feature already popular on radio.

55. Picture Parade scripts, no.1. WAC.

56. Kine Weekly, 28 February, 1952, p.5.

57. 3 May, 1935. CEA to BBC. R51/173/1.

58. 15 March, 1946. Peter Eton to Alan Curthoys. R19/914/1.

59. 19 March 1951. Cecil McGivern to John Dennett. T6/104/1.

60. '1952 Viewer Survey', folder R9/4.

61. 30 January, 1952. Small to McGivern. 'Notes on conversation about telefilm repeat of "Current Release". T6/104/2.

62. 31 January, 1952. Small to McGivern. T6/104/2.

63. As described in the Picturegoer, 9 February, 1952, p.3.

Chapter Three

Current Release: Text and Audience

Using the scripts and other archival sources, this chapter offers a textual analysis of *Current Release* in terms of aesthetics, technology and developing modes of televisual address - how the programme was organised and designed, and how it brought the cinema into the home. This is placed within the context of the discursive negotiations between the BBC and the film industry and their role in shaping the form, content and address of the series. The circulation of the programme is then explored from a different perspective - that of audience reception - which situates *Current Release* within an economy of early televiewing and shifting attitudes towards cinemagoing. This emphasises how the early cinema programme occupied what was very much a transitional space in terms of changing leisure patterns, and the consumption of screen entertainment in public and private spaces.

The Programme Text

Current Release began on 17 January 1952 and ran on a fortnightly basis for thirty-two editions. It was initially scheduled on Thursday evenings at 9:15pm, and this was not a peak day within the weekly television schedule. Various forms of drama dominated the BBC's output at this time (see Caughie, 1991a), and on Thursday night, the Sunday play was repeated by way of a second live performance. As a result, Thursday evenings were expected to command the smallest audience of the week. While the popularity of *Current Release* was to prompt a shift in its scheduled slot, this indicated that the BBC also had some uncertainty as to how this essentially 'experimental' venture would be received. With regard to the film industry, however, the reaction to the first edition was largely favourable. *Kine Weekly* described the series as a 'courageous experiment and a co-operative venture that stands to benefit both [media]',[1] while *Today's Cinema* reported that 'General industry response was that it was put over well and would prove an incentive for... people to visit the cinema'.[2] The interest of the film industry was in fact so keen that, for the first few editions of *Current Release*, the trade press published detailed analyses of the programme. This discursive context doesn't simply represent a form of commentary *on* the programme (which can be used as a historical source), as it also contributed to, and shaped its development. When *Current Release* began the BBC explained that they 'would be sorry if trade paper criticisms were in any way to be restricted as the BBC had no objections [to them] whatsoever'.[3] The BBC aimed to collect the press responses to all of its output, and when it came to the cinema programme, this included the film trade papers. In conjunction with the views expressed at meetings, the trade press was part of the 'dialogue' between the film industry and the BBC, and was openly understood as such. This was partly because, unlike the film-orientated programmes in the American context where the film companies were involved in their production, the BBC retained full editorial control over the programme. Nevertheless, there was an awareness that the trade press discourse could (and should) exert a shaping influence on the series, and it represented a dynamic site of debate where the cinema programme was concerned throughout the decade.

The first factor to emphasise is that *Current Release* was transmitted live, and this exerted a shaping influence on its textual form. While television's use of film was expanding at this time (Barr, 1996, p.55), television was still perceived to be essentially a live medium. As many critics have stressed, 'liveness' was considered to be an aesthetic virtue of the medium, rather than a technological necessity (Caughie, 1991a), and this shaped institutional and cultural perceptions regarding the use of filmed material. As John Swift's *Adventure in Vision* explained, 'The transmission of action recorded on film is not television in its true meaning' (1950, p.185), and he perceived that only the BBC newsreel justified its use. Jan Bussell's *The Art of Television* agreed that film should be used as little as possible as 'to borrow, is a confession of weakness' (1952, p.55), and comparable to Bussell's dislike of the 'canned deadness of a piece of film' (p.55), the producer of *Current Release* felt that filming it would 'impose so many limiting factors on the style and content [of the programme] as to expose it to the danger of repetitiousness - or even dullness'.[4] A further disadvantage was that of cost as to telefilm the series would cost almost three times as much as a live broadcast. The BBC estimated that produced live, *Current Release* would cost approximately £110 per edition - against £310 per programme if telefilmed. Along with programmes such as panel games and popular drama, studio-based magazine shows represented a form of low-cost programming for the BBC (and later ITV), and this was a key part of their appeal for broadcasters (Thumim, 2002b, p.212). If telefilmed, *Current Release* would also have taken longer to produce, making a fortnightly feature impossible. The film industry had favoured a *weekly* series, and were thus quite resistant to a fortnightly feature. A range of aesthetic and economic factors shaped the decision to broadcast *Current Release* live and at least in terms of the BBC, this remained the preferred method of transmitting the cinema programme until into the 1960s.

Within the context of early television, critics have emphasised the degree to which, due to conceptions of its live aesthetic and 'intimate' domestic address, the screening of feature films represented not only the 'institutionally unattainable, but also the aesthetically undesirable' (Caughie, 1991a: 32). Yet it is clear that, by its very definition, the cinema programme involved the combination of these apparently 'oppositional' aesthetic forms: live television and film. But a key difference between the screening of the feature film and the cinema programme is that the cinema programme endeavours to construct a cinematic 'world', a cinematic *culture* within which the films can be situated. It is here that an aesthetic of 'liveness' appeared to play a central role in the programmes from the 1950s.

Although the programmes themselves no longer survive, it is important, as far as possible, to 'reconstruct' the form, viewer address and visual aesthetic of *Current Release*. While the scripts may place the emphasis on the verbal aspects of the programme, they also offer insight into its general 'look' and its set. The set invited the viewer to engage with the text in particular ways and, crucially, it became a site of conflict in the development of the early cinema programme. As an area of contestation between the BBC, the film industry and its viewers, it suggested the ways in which the cinema programme was situated within conceptions of television's developing modes of address (and in some respects, the instabilities which surrounded these). Yet more specifically in terms of the genre, it suggests the ways in which the Corporation envisaged how the cinema might be rendered 'televisual' for consumption through the live, direct address of the medium and in the domestic sphere of the home. The first two editions are revealing in this respect as they were substantially different to those that followed.

'The "BBC" End of the Programme': Constructing a Home for the Cinema

Current Release initially made use of three different sets which simulated a 'cutting room', a 'reception room' and an 'office'.[5] The office was constructed of a desk, a typist's chair, filing cabinet, hat stand and what were described as several 'glossy film portraits' (depicting 'current stars'). The cutting room housed film reels, an editing suite and a projector, while the reception room, in which stars and directors were interviewed, offered a comfortable surround complete with a 'Chesterfield suite and Chippendale chairs'. Spread out on the coffee table was a selection of film trade papers, and this was situated adjacent to a sideboard sporting whiskey and gin, lamps, flowers and what are described as various 'objets d'art'. The first thing to note is that these sets were entirely the idea of the BBC, and there appeared to have been a genuine delight in exploring the visual possibilities of television in ways that radio had not required. However, from the BBC's perspective they were by no means simply decorative. The producer was quite specific about the functions of the set and explained them in detail. Firstly, he wanted to use 'actors to simulate film technicians, officials, secretaries, doormen etc', in order to create an 'atmospheric type of linking' for the studio material. Such an 'atmosphere' he suggested, would afford 'interest, variety and verisimilitude to the general magazine content of the series'.[6] The use of the term 'verisimilitude' here is particularly revealing as to the producer's conception of the programme's set - how he intended it to be 'like', or 'similar' to, a *real* office or cutting room. In terms of fictional texts, verisimilitude is a term used in film, television (and previously literary) studies which is taken to mean 'probable' or 'likely' - what we are willing to accept as 'realistic'. As Richard Maltby explains: 'Verisimilitude implies probability or plausibility, less a direct relation to the "real" than a suggestion of what is appropriate' (1995, p.109). This appears to be partly the context within which the producer is using the term. In a programme dealing with films and film culture, it was considered 'appropriate', *natural* even, that it should present a space peppered with iconography and people from the world of the cinema. If verisimilitude serves to support and shape a construction of the 'real', then the producer's approach to the set reflects on an early conception of television *realism*. This is not so much at the level of aesthetic form, but in terms of how television was seeking to connect with the world 'out there', in this case, the world of the film industry and the cinema.

This was not the only function of the set. The producer continued:

> The intention... is to create a local habitation for the programme so that, without long explanation or apology, the narrator can, as the sense of the programme may require, take up various aspects of filmmaking and exhibiting and enter into a discussion and demonstration of such aspects with experts and others. This is the "BBC" end of the programme... it should have a high interest appeal to viewers and not merely be a flabby framework for the expertise of ... filmmakers.[7]

As this suggests, the set also played a practical role in terms projecting what were perceived as the more 'informative' aspects of the film programme. The context of the set is quite specifically aligned with 'the BBC end of the programme' and thus, from the Corporation's point of view, it had potential links with public service. It also suggests, however, how *Current Release* hovered on the boundary between the fictional and the non-fictional. The set was to facilitate the educational aims of the programme, but *not* at the expense of the fictional world it aimed to construct. According to the producer, it was undesirable that the presenter should be seen to explain why certain aspects were included; these were to arise quite naturally from the programme's *mise en scene*. Such a conception indicates that Fitzgerald was not to step out of the 'local habitation' and almost fictional context in which he was placed, as this would destroy its illusion of reality. The cutting room, for example, was to be used to demonstrate particular filmmaking techniques, but it was not to be apparent that this was its only function. It was to be offered to the viewer as a *real* cutting room and thus as integral to the programme's aesthetic environment. *Current Release* was to present an almost self-enclosed and pre-existing world in which the viewer was invited to *believe*.

At the beginning of the first edition Fitzgerald is seen in the cutting room with his back to the camera, speaking on the telephone: 'I know, I know. There's no time to get back to the office now... I'll start the programme here...' He then turns to face the camera and continues a conversation with his secretary: 'It'll be alright Susan... Yes. You bring my notes up here right away...' After establishing this environment Fitzgerald turns to address the viewer, inviting them into the fictional world of which he is part: 'Hello. Sorry about that. I hope you're going to like this programme all about films and film folk'.[8] (Notably, the use of the term 'folk' was informal, indicating the BBC's desire to convey a relaxed, informal and accessible approach). Fitzgerald then explains that the films have been carefully selected and he attempts to locate them as the camera tracks back to reveal a stack of film cans: 'Yes - here we are. *The Magic Box* [John Boulting, 1951], *Cyrano de Bergerac* [Stanley Kramer, 1950], *Encore* [Pat Jackson/ Anthony Pelissier/ Harold French, 1951], *I Want You* [Mark Robson, 1951], *Lady Godiva Rides Again* [Frank Launder, 1951] and *Where no Vultures Fly* [Harry Watt, 1951]'.

Immediately following this reference to the films, Fitzgerald deliberately emphasises that *Current Release* is not simply a series of 'trailers':

> This programme, by the way, besides telling you about current films will also tell you about the business of making films... You'll be seeing quite a lot of what goes on in this room, for instance. This is a cutting room. It's a place where all the hundreds of shots... are cut and joined to make the continuous reels you see on the cinema screen.

As this suggests, the cutting room was indeed presented as real, as though the cameras were simply peeping in on its typical daily work. 'Susan' is Fitzgerald's secretary and she enters the set to bring him his papers: 'Here are your notes, John. Don't muss [sic] them up. They're in the proper order. *The Magic Box* on top'. These 'notes' were in fact Fitzgerald's script. Without the benefits of an autocue, scripts were often used on television at this time and with regard to the BBC's cinema programme, continued to be visible for much of the decade. Fitzgerald then instructs Susan to take the first reel to 'Charlie' (the projectionist), as 'he'll want to get it laced up. Tell him he can send for the others as soon as he likes'.

Firstly, Fitzgerald's dialogue is indicative of the 'spontaneous' and almost chaotic atmosphere which characterised the first edition (and, to a lesser degree, the second). What seems crucial here is that this deliberately capitalised on the liveness of television and the relations this constructed between viewer and screen. Live television truly had an air of unpredictability and *Current Release* exploited this while exploring its possibilities within the programme's 'fictional' world. There was apparently no time for Fitzgerald to reach his office, he had to 'improvise' by beginning the programme in the cutting room. This atmosphere extended to other aspects of the programme such as the arrival of the guests. Later in the episode Susan interjects: 'John, sorry to interrupt. [The director] Harry Watt's just come in. I've put him in the drawing room and given him a drink.' Fitzgerald enquires 'Any sign of [the interviewer] John Parsons yet?', to which Susan replies, 'He's been in, I believe, but I've lost him for the moment'. A similar air of unpredictability characterised episode two when the young British actress

Eileen Moore arrives to discuss her recent film, *Mr. Denning Drives North* (Anthony Kimmins, 1951). As she appears Fitzgerald exclaims, 'Well - so you've made it. Good! Come on in'. Rather than waiting off-screen ready for her cue, Fitzgerald implies that the appearance of Eileen Moore was in fact subject to chance.

While this atmosphere could equally be created in a filmed programme, part of its meaning and effect relied upon its *live* transmission, and the expectations surrounding this aesthetic. The producer was keen to capitalise on its unpredictability - the fact that items could go wrong on live television (and frequently did), or that timing schedules would not go to plan. Caughie describes how the immediate years of post-war television were still driven by 'the enthusiasm of the amateur inherited from the pre-war pioneers. The whole discourse of production, the celebration of disaster, the informal working relations, the try-outs, carried something of a "wizard prang" about it...' (1991a, p.40). Here, however, this rhetoric was to be almost 'fictionalised' and made into an entertaining spectacle in itself. Its most important effect was reinforcing the impression that the audience were truly peeping in on the environment and almost catching its inhabitants 'off guard' - a bustling world which had an independent existence outside the time the programme was on air. While this will be discussed in detail in relation to the 'behind-the-scenes' footage in Chapter Six, what is intriguing here is that there is an attempt to generate a sense of a busy, thriving film industry - the very concept that television itself was on the brink of helping to destabilise.

Film programmes on radio could also offer a playfully theatrical approach in which speakers adopted 'character' roles. As indicated earlier, in the late 1940s, speakers in radio's *Picture Parade* played the parts of 'Miss Film Fan' (a working 'shop-girl' with a passion for the 'flicks') and 'Mr. *Picturegoer*' (a middle-class and supposedly more 'discerning' spectator) in an attempt to represent the BBC's conception of a broad listening audience. However, while these speakers were playing roles, characters even, it was not the same as creating a 'fictionalised' world. It was also possible that what was acceptable on radio was not necessarily successful in vision. While *Current Release* needs to be understood within the context of radio's coverage of film it is clear that, in many ways, a televised cinema programme was quite a new proposition. The coverage of film on radio had exploited the specificities of the medium, investing in the musical and aural aspects of film. *Current Release* was exploring, from a tentative and experimental perspective, what was most appropriate for television. From a contemporary point of view, what may appear to be its somewhat 'curious' construction equally speaks to the fact that, in comparison with film for example, there is a less elaborated historical discussion of how television 'grammar' - its codes and conventions - developed in these early years (Caughie, 2000). This is particularly so in relation to forms of non-fiction programming, and especially magazine-style fare (which notably remain one of the most neglected areas of television analysis in general). In trying to 'reconstruct' how the programme was conceived and the decisions and indecisions which shaped its early development, what may now appear to be somewhat bewildering (while clearly fascinating) choices perhaps reflect Janet Thumim's point that the dilemma for researchers can often be 'in *seeing*... what was "on television" at all in the 1950s...' (2002a: 2).

Television Reveals the 'Tricks of Filmmaking'

The opening sequence of *Current Release* not only raises the issue of the 'unpredictable chaos' of live television and its relation to the construction of a cinematic world. In following the films from their cans to the projection room, it was also self-reflexive in exposing the production process behind the programme, part of a broader strategy in which the aesthetic and technological differences between film and television were constantly foregrounded in a number of ways. While drawing on cinematic iconography in imaging the stacks of film cans and the reels of celluloid, this cannot be viewed as fictional in the same way as the set. The sequences really *were* shown by an ordinary cinematograph projector (combined with a television transmitter) to create what was known as a 'Telecine' (Barr, 1996, p.55). While emphasising a convergence between the media - films can be transmitted on television - the visibility of this technology equally served to differentiate them. With its hazy beam of light and rolling reels of film, the projector is associated with the specifically cinematic. Within the context of *Current Release*, and again feeding back into the importance of television's live aesthetic, it emphasises how film cannot easily be assimilated into television's textual flow, it requires the use of appropriate and 'special' equipment. In the foregrounding of the production process, this also suggested that television technology was of interest in itself. When *Current Release* was on air the BBC broadcast a programme called *How Television Works* and the *Radio Times* billed 'a film showing how the television cameras bring the picture to the screen'.[9] While intending to perform an informative function, this also indicates the perception that the audience are interested in the workings of a new and 'mysterious' technology. In terms of *Current Release*, it was television's ability to transmit films which was considered to be of interest, and this was repeated in other cinema programmes in these early years. This foregrounding of *process* rather than content carries the hallmark of a new technology, and such self-consciousness receded once the series got under way.

As Fitzgerald's opening comments make clear, interesting the viewer in the processes of production was part of the programme's aim, although this was in terms of the feature film. The cutting room demonstration took place in episode two and enlightened viewers on the process of editing, as well as the relations between editing and narrative. Howard Thomas, the Producer-in-Chief at Associated British Pathé, had previously contacted McGivern to explain:

One thought [for a television programme] I would like to put before you is a discussion on the art of film editing. Editing is one of the main points of difference between live television and filmmaking and I am sure viewers would be interested [in this]... We always find that visitors to our organisation ... are fascinated ... by the movieolas and editolas.[10]

Shortly after this the BBC produced a series entitled *Around the Film Studios* which 'showed viewers how both feature films and newsreels were made',[11] and *Current Release* continued this perspective. In defining television as a live medium, Thomas clearly perceives the interest of this subject to emerge from the technological and aesthetic *differences* between film and television. As the 1950s progressed, television's use of film steadily increased, encouraging a blurring of the perceived oppositions between cinema and television, not only on the level of what Barr terms the 'institutional frameworks, economics and consumption ... but also of "language" and aesthetics' (Barr, 1996, p.50). By the mid-1950s it was possible to shoot the television image on film, and with the introduction of the Ampex machine in 1958, it was possible to record the television image on videotape. (Editing the tape-recorded image had to wait until the early 1960s) (Barr, 1996, p.58). In 1948, however, it is not surprising that Thomas speaks little of these possibilities nor that, in 1952, it should be seen to represent a point of interest for the viewers of *Current Release*. This showcasing of film in the first edition of the cinema programme - whether the emphasis on the film reels, the foregrounding of the projector or the demonstration of film editing - all points to a self-consciousness about the introduction of film to the home. It rehearses at a visual level the domestication of film culture, and the novelty of what this might entail. In doing so, *Current Release* was emphasising what, at the time, were understood to be the key technological and aesthetic differences between film and television.

Important here is that the programme's coverage of an 'inside' view of filmmaking seemed to offer a new perspective to viewers, as indicated by the response of both the trade press and the audience. After the first edition on editing the trade criticised what *Kine 'Weekly* described as 'the behind the scenes tricks of filmmaking'.[12] The report commented that the demonstration was done well, 'but we may ask whether... it is a good thing to take kinemagoers behind the scenes?'[13] In episode three, *Current Release* previewed the film *I Miserabili* (Riccardo Freda, 1947), starring Valentina Cortesa and Gino Cervi. This was an Italian remake of *Les Miserables*, and for its British and American release it had been dubbed in English. The BBC invited film critics C. A. Lejeune and Campbell Dixon to discuss the implications of dubbing from an 'artistic' point of view, and posed the question: 'Is it merely a form of reproduction or is it, in fact, a form of misrepresentation?'[14] The programme also demonstrated the process of dubbing and *Kine Weekly*'s critic again complained that he could not 'see the advantage of showing the public behind-the-scenes tricks of filmmaking... it was surely wrong to reveal the dubbing method as far as *Les Miserables* was concerned'.[15] These responses seem to indicate a desire to retain an illusionist discourse around the cinematic product. The comment that it is 'surely wrong' to reveal the production process suggests a concern that this may reduce the public's enjoyment (and thus interest) in film - it might prevent them from 'immersing' themselves in the fiction. The *Daily Film Renter* simply considered that the cutting room demonstration seemed 'rather too elementary to give much added interest to the programme',[16] yet this obscures the fact that it offered what was essentially a new perspective to the audience. A BBC Viewer Research report specifically emphasised that viewers 'found the glimpse of the cutting room very interesting',[17] and twelve months later a viewer was able to recall the edition as one that he and his wife were 'particularly interested in...[It showed] the "inside" workings of the film industry [and]... was very educative... more of this type [of material] would stimulate picturegoing interest'.[18] While many viewers criticised what they perceived to be the programme's 'make-believe' construction of a 'cinematic' world, the glimpse 'behind-the-scenes' was taken to be factual and *authentic*, despite the fact that (taking place in the 'cutting room'), it was no less constructed than Susan's secretarial role or Fitzgerald's busy 'office'.

While we are familiar with the 'behind-the-scenes' perspective today, this now tends to refer to what happens on the film set, how scenes are filmed or stunts performed and so forth. Television has played a central role in developing this perspective, and its emergence is later discussed in relation to *Picture Parade* and *Film Fanfare* in Chapter Six. *Current Release* was concerned, in the first two editions at least, with the more technical side of filmmaking, and in terms of constructing a possible framework of interpretation, it is worth acknowledging where such a perspective was likely to be found, and equally, where it was not. Particularly in the 1930s and 1940s, the BFI publication *Sight and Sound* tended to focus on technical issues and, in contrast to later years, its adverts were primarily aimed at the film trade (dominated, for example, by adverts for projector equipment). A further source is the *Penguin Film Review* (1946-49) which explored a range of perspectives on British, American, and other national cinemas from the industrial to the aesthetic. As part of this, it sometimes examined the more technical side of filmmaking and the 'craft' involved, such as the article by John Shearman entitled: 'Who are those Technicians?' (1946) which explored the complex craft of a variety of roles which take place behind the camera. Yet although succeeded by *The Cinema*, the *Penguin Film Review* ceased publication by 1950 and suffice it to say that, like *Sight and Sound* at this time, it was not a 'light read' for the average cinemagoer. The 1940s had nevertheless witnessed the publication growth of an increasing number of books on filmmaking (Richards, 2000, p.24). Written in 1951, Roger Manvell's *A Seat in the Cinema* (1951) provides an interesting example. The inside cover reads thus:

For the cinemagoer who wants to know more of the 'inside story' of filming, and for those who wish to attain a deeper appreciation of the film as an art form - this book will prove invaluable. Here the reader will find an account of all the work and skill that goes toward the making of a film - the parts played by the many technicians and specialists.

Among other things, the book described the stage by stage process of scripting, shooting and editing a film, and the breakdown of labour this involved. It also contained stills depicting these stages, including, for example, a picture of 'a dubbing session in process' (p.80). Again, however, it is questionable to whom such a book would appeal, and thus the type of circulation it would have received.

Perhaps most importantly, the technical aspects of filmmaking were not generally covered by *Picturegoer* fan magazine. In 1953 one respondent specifically complained that 'There never seem to be any articles on the technical side of film making.... in *Picturegoer*. The more one knows of a subject, the more fascinating it becomes',[19] although not everyone agreed with this perspective. Reminding us of *Kine Weekly*'s concerns, another reader commented: 'Take away the seventh veil from the dancing girl and the illusion is shattered. Tell us often enough how pictures are made and a bigger illusion has gone. The less people know of the technique the better'.[20] The magazine explained that 'most producers' agreed, 'preferring to concentrate on what is on the screen and not behind it'. These letters are interesting not because they are statistically reliable in offering opinions on the subject, but because they indicate that this was not a perspective really covered by *Picturegoer* at all (nor its rival magazine *Pictureshow*).

It is possible that *Current Release* represented one of the first examples of unveiling the 'behind-the-scenes' perspective on a wider scale. Perhaps the reasons the BBC were keen to include such a perspective are indicated by the quote from Roger Manvell above. According to Manvell, his book is for 'the cinemagoer who wants to know more of the "inside story" of filming, and for those who wish to attain a deeper appreciation of the film as an art form'. While he separates the two conceptions here in so far as finding out about the 'inside story' of filmmaking need not lead to a 'deeper appreciation' of its 'art form', it is clearly perceived as a possible link - a logic which perhaps prompted the BBC's desire to explore this field. This again suggests the ways in which the cinema programme was developed under an ethos of public service, and how this was structured by the recurrent concern to balance this agenda with the programme's precarious relations with 'promotion'.

'Gendering' *Current Release*?

As well as the 'behind-the-scenes' perspective and the programme's fictional world, a further issue raised by the opening sequence is that of Susan, or more specifically, her relationship to questions of a gendered address in the programme. It is widely perceived that, after the temporary liberations during wartime, the 1950s were characterised by an attempt to impart home-making as a career to women of all social classes, but this picture has since been complicated. It was only women's full-time employment which declined after the war, as their part-time work actually increased, leading to a rise in the number of married women in employment (Kingsley Kent, 1999, p.320). Rather than women simply conforming to the domestic ideal, the issue of the working woman was a hotly debated site of ideological struggle. As Janet Thumim argues in her analysis of the representation of 'Women at Work' in popular British television drama in the late 1950s, despite 'the media emphasis on stories of *exceptional* "career women", experience showed that women's economic contribution to family income was vital in securing the newly available consumer goods, just as their labour was vital in producing them' (2002b, p.214). Although the wider detail of these social and cultural shifts are beyond the scope of this analysis, within this context, the BBC's decision to include the figure of the 'working woman' in a popular cinema programme is an intriguing one - and it was a strategy which continued with *Picture Parade* until the late 1950s. Recall that, according to the producer of the programme, the use of a secretary was intended to contribute to the 'verisimilitude' of the set - effectively its realism and plausibility. If Susan was to offer an image so 'natural' that it would appear 'realistic', she perhaps offered a specific representation of the working woman, one that was socially acceptable. A clerical job was indeed one of limited number of options open to women, and was perhaps more attractive than the distributive trades or factory work in which they were regularly employed. This was particularly so for young girls who entered employment straight from school, and the job was cited as a 'stop-gap' before 'a young woman began her true "career" as a wife and mother' (Summerfield, 1994, p.66). Although it is difficult to tell from the script, there is the sense in which Susan seems to be a *young* working woman, and her tasks were those we might expect of a secretary. She brought Fitzgerald his important papers and played hostess to the guests, although it would be misleading to suggest that Susan was simply a subservient figure or part of the 'backdrop' of the set (as the producer's conception of her role implies). As we have seen, she instructed Fitzgerald to keep his papers in order and often prompted his lines, and her comments also occasionally extended to the films themselves in ways which significantly foregrounded issues of gender. For example, she was seen to make such remarks as 'You won't say it's a "women's picture" will you John?', or on one occasion she asked whether a film was simply 'propaganda to make us working girls more contented with out lot?'[21] Although a female presence in the cinema programme was marginalised when it came to selecting a presenter for the series, Susan's references to women here - both as an audience for the cinema and as 'working girls' - allude to the possibility of attracting a female audience for *Current Release*.

Television's entry into the British home will be discussed in connection with the reception of *Current Release*, but women have been seen as central to the insertion of television into domestic spaces and routines, at least with regard to the American context (Spigel, 1992, 1996). Women have historically also been seen as occupying a central role in the mass audience for the cinema. Much of the work surrounding this issue has concerned itself with the relations between women, cinema and consumerism - aspects of which are addressed later in the book in terms of the television's showcasing of star fashions at film premieres (Chapter Six). Largely in relation to Hollywood cinema, critics have considered mediating factors such as tie-ins developed through fan magazines and other intertexts which positioned the female spectator as consumer (La Place, 1987, Doane, 1989, Stacey, 1994, Klinger, 1994). Historical research in the American context has also indicated ways in which this address was articulated by television in the 1950s in terms of its increasing domestication of film culture. As Denise Mann argues in her discussion of Hollywood stars in early television variety shows, 'The fan and mass-circulation magazines surrounding Hollywood and its star system marked women spectators as an economically and socially viable group - one which the broadcast industry sought to incorporate as an audience for its own mass-media forms' (1992, p.42). Mann's analysis here raises also raises the point as to whether women have historically been perceived to be more interested in the *intertextual* circulation of cinema - coverage of and interviews with the stars, for example. The address of the British fan magazine *Picturegoer,* and certainly its advertisements, certainly always seemed more inclined toward a female reader. The composition of the cinema audience was subject to fluctuation (see Harper and Porter, 1999), so this is not the same as saying that women were more interested in filmgoing than men. But it is significant here in so far as the cinema programme was to function within this intertextual, and now domestic, space. Women's programmes on radio (such as *Woman's Hour*) had long since included coverage of film culture, and it seems logical that these relations would continue with cinema's further penetration of the home via television.

Nevertheless, this suggestion of women's more pronounced interest in the intertextual construction of the cinema is speculative. Furthermore, in considering the cinema programme, there is a potential danger in over-emphasising the perception of women as the primary consumers for television. The promotion of television to American women was significantly shaped by the institutional basis of US broadcasting, its commercial nature, and the fact that women were the primary target for advertisers. While the perceived centrality of women as an audience for television is certainly also shaped by its domestic location, institutional differences between British and American broadcasting make a simple appropriation of these arguments problematic. In terms of the British context, Joy Leman (1987) has considered the development of 'women's programmes' on 1950s television (the first beginning in 1947 with the BBC's *Designed for Women*), but these were scheduled during the day when we would expect women to be regarded as the primary consumers of television. The main cinema programmes were all scheduled during the evening when it is less simple to make such claims. It is clear that throughout the period covered in this book, the cinema programme aimed for a very broad appeal: seeking to attract men and women, the young and old. This hardly seems surprising given that the cinema had traditionally been a popular mass interest, and television was in many ways now courting the family audience which had once been so loyal to the cinema. In *Current Release*, Fitzgerald specifically invoked a general address when he explained how the BBC 'try to include something for everyone in this programme',[22] yet the appearance of Susan and her occasional references to women's filmgoing tastes nevertheless suggests an attempt to address *women* within this broad appeal. She also recalls the address of radio's *Picture Parade* (1949) with the use of the speaker 'Miss Film Fan': 'I'm what so many of you are, a film fan. I work hard all day in my office. I go to the flicks to be entertained - and I most certainly know what I like'.[23] 'Miss Film Fan' was envisaged by the BBC as a conception of the young, (working class?), female cinemagoer, an example of what was perhaps intended to suggest a modern working woman. This is clearly conceived as an address toward women ('I'm what so many of *you* are'), and it is possible to see a connection with 'Susan' here. She was perhaps intended to represent a *modern* working woman who even enjoyed a banter with her 'boss', with what the BBC perceived to be a likely interest in film culture. Furthermore, in itself related to women's positioning as an audience for the intertextual circulation of the cinema discussed above, the conception of the 'fan' here as female is significant. Perhaps the selection of a male presenter for *Current Release* was an attempt to 'masculinise' the address of the cinema programme, and to ward off what were seen as the gendered, female connotations of 'fandom' and frivolity with a rhetoric of casual authority. At the same time, evidence will later indicate how Fitzgerald was very popular with female viewers.

Susan's (so-called) 'interruptions' of Fitzgerald's dialogue attracted many complaints from both the trade press and the viewers. She was seen as unsuitable, frivolous and distracting, qualities associated not only with 'the feminine', but with the BBC's prejudices toward women in terms of their role in developing television's modes of address (Thumim, 1998, p.102). Maurice Wiggin of the *Sunday Times* spoke of the 'mumbo-jumbo of the presentation whereby we are invited to pretend that we are in some sort of flash office, complete with brisk young executive and decorative secretary',[24] but it seemed to be precisely the point that Susan *wasn't* simply 'decorative' which prompted the complaints. She did not just contribute to the 'verisimilitude' of the programme as the producer had intended but, as the responses suggest, she could also be seen to undermine Fitzgerald's role. Perhaps more crucially, this may emphasise that while women may have been acknowledged as central to television's insertion into domestic space and daily routines, 'The female presence on screen was carefully contained' (Thumim, 1998, p.91).

Maurice Wiggin's reference to *Current Release* in terms of a 'pretence' epitomised a view expressed more widely by both the film industry and the viewers. According to the Viewer Research reports, the audience were also confused by what the set, in its entirety, actually proposed to be. The inclusion of a cutting room appeared to indicate that it was a place where films were made - the production unit at a studio perhaps. The 'office' was possibly where the studio administration occurred. Yet if it was a production unit, a film studio, then why did it undertake the regular job of previewing films for the *television* audience? As this suggests, it required the viewer to suspend their disbelief, and this was in keeping with the demands of a fictional genre. Evidence suggests that viewers particularly disliked the artificiality of the performance, and the way in which this shaped Fitzgerald's role and audience address. In contrast to the responses to Susan, reaction to John Fitzgerald was very favourable, yet viewers criticised what they called the 'unnatural' or 'artificially casual' way he introduced the programme. The word 'casual' was used here not to refer to an easy-going or relaxed approach, as we might understand the term today, but to the pretence that Fitzgerald was in a busy working environment which forced him to 'improvise' his lines. In general, the viewers and the film industry held quite the opposite conception to the BBC. Rather than 'realistic', they perceived the set to be phoney, inauthentic, and generally *in*appropriate.

These reactions to *Current Release* were essentially debates about how television should address its audience, and how this might be mapped across different types of genre - something clearly unstable and experimental at the time. It was not only the film industry and the press which offered a negative reaction, but also the Programme Controller, Cecil McGivern. Subsequent editions were presented solely from the reception room set, containing the same furniture and objects as before. These reactions all served to steer *Current Release* on a different course than that originally conceived - although, as the later *Picture Parade* and *Film Fanfare* will make clear, this diversion was not necessarily permanent. While by no means identical to *Current Release*, these programmes suggested that a 'cinematic world' was still an appealing concept and in 1956 it re-emerged with a startling enthusiasm and vigour.

However, what it seems to reflect on here, is that there was no simple understanding of a division between fiction and non-fiction programming (and of course their formal and aesthetic strategies have always been, and continue to be, blurred). More specifically, it suggests an experimentation and instability over where to situate the cinema within this space. Of considerable interest here is that when the producer subsequently attempted to play with the set (by adding some form of 'window'), the Programme Controller articulated his changed conception of *Current Release*. He informed the producer that 'the "set" was wrong. Silly, in fact, and offends correct documentary thought. You used a window set in a realistic programme. Worse still, it was daylight lit and the transmission was taking place at a time of darkness'.[25] It is not so much the window set which is of importance here, but the fact that *Current Release* is described as a 'realistic' programme, aligned with the visual aesthetic of the 'documentary'. While clearly being used in a different context than is the case today, BBC Viewer Research Bulletins classified *Current Release* not under 'light entertainment' (as with general magazine programmes such as *Picture Page* for example), but as a form of 'documentary' programme. Although subsequent cinema programmes were keen to play with the construction of 'appropriate' cinematic spaces in terms of their sets, this nevertheless marked the time at which the genre was to be conceived as essentially non-fictional and 'realistic' - separate from the 'frivolity' of 'make-believe'.

Current Release: Talking Film

In comparison with later programmes, *Current Release* was more straightforward in its format, combined of film previews and interviews. This also offered a contrast to the film programmes on radio given that they were much more of a bricolage of film culture old and new. The film industry had resisted this approach with *Current Release* by arguing that (in keeping with its title), it should deal only in contemporary film culture - thus maximising, of course, its promotional appeal. Given their concerns over this issue, the BBC had agreed to this rather reluctantly. It is clear from the evidence that they were initially keen to produce more of a film magazine, similar in scope to that of radio's *Picture Parade* or *Filmtime*.

In its attempt to appeal to a broad audience, *Current Release* featured a range of genres in each edition,[26] and the scripts indicate how discursive conventions for presenting the films were quickly established. After outlining the plot and mentioning the stars, Fitzgerald often elaborated further on the film after the excerpt was screened, summarising particular dramatic, amusing or aesthetic points of appeal. He appeared, for example, to take a particularly keen interest in the scenery and settings of films. In discussing *Never Take No For an Answer* (Maurice Cloche/ Ralph Smart, 1951), for example, he suggested that 'This film would almost stand up on its unusual and beautiful backgrounds in Assisi and Rome'.[27] When previewing *Cry the Beloved Country* (Zoltan Korda, 1951) Fitzgerald was keen to enthuse about its South African setting and insisted that 'There's no mistaking the authenticity of the backgrounds, and certainly no avoiding the eloquent contrasts between the fruitful valleys and the shanty towns'.[28] Introduced into the conversation in a relaxed, informal manner, this was perhaps an attempt to encourage viewers to look more closely at the visual aesthetics or art of film beyond the plot, the action, or the stars. Fitzgerald's commentary was certainly not deliberately complex and it was shaped by a desire to meet the viewer 'on their own ground', or rather the BBC's perception of what that ground might be.

The BBC was certainly mindful of the need to make their first TV cinema programme popular in tone. On occasions when Fitzgerald's dialogue was more complex, it was self-consciously mixed with a more colloquial approach. In episode two, for example, Fitzgerald introduced the science-fiction film *The Day the Earth Stood Still* (Robert Wise, 1951):

> This is one of the best of the current cycle of pictures dealing in one way or another with interplanetary travel and atomic energy. One important distinction between this picture and [the]... others... is that... it lays rather less emphasis on the mechanistic angle - you know, what you might call *'knobs and test tubes'* - than on some of the social and political implications of the liberated atom [my italics].

It has often been suggested that, particularly when compared to ITV, the BBC's presenters were formal, didactic, and patronising in tone. Peter Black describes how 'ITV forged a friendly style of presentation which the BBC had always found elusive' (1977, p.116) while Bernard Sendall insists that ITV presenters 'set up a personal relationship with the viewer', while the BBC's were purely 'mouthpieces for the Corporation' (1982, p.325). Given that so little of early programming has survived, these generalisations are open to question, and are related to wider aspects of broadcasting history and its tendency to polarise the 'stuffy', 'paternalistic' and 'didactic' BBC with the new energetic populism of ITV. These perceptions also raise much broader issues about the relationship between 'macro-overviews' of broadcasting history (Jacobs, 2000: 9) and the more local analyses of specific genres or texts - and indeed the extent to which these may not always offer a 'neat' fit. It is true to suggest that, despite the BBC's desire to use a 'representative of the ordinary filmgoer' to present their first TV cinema programme, as well as the considerable emphasis on the importance of an informal and casual mode of address, Fitzgerald's accent conveyed very middle-class tones.[29] Yet these arguably epitomised televisual speech in this period, and there are no clear-cut distinctions in this respect between the BBC and ITV cinema programmes (not least of all because certain presenters were moved between the two channels).

Fitzgerald's commentary was monitored very carefully by the BBC, particularly with respect to the acute concern over 'advertising'. The boundaries within which Fitzgerald's commentary was to operate were highlighted in the discussions over *Current Release* when the film industry submitted their pilot script to the BBC. McGivern had commented that 'The style of the commentary is not good.... It would seem pretty cheap coming out of a television set into the comparative intimacy of people's homes. It has the facile atmosphere of the small-type showman or the "stage" commercial traveller'.[30] His response here emphasises two points. Firstly, Fitzgerald's discussion of film had to be carefully attuned to the *domestic* context in which it was received, and prefiguring the concerns later raised by ITV, this was a sphere in which advertising was seen to be particularly inappropriate. Secondly, McGivern's comment recalls a very early stage in the cinema's own development when films were indeed presented by a travelling 'showman'. While this issue cannot fully be explored here, it is interesting to speculate whether the film programme may have a precursor in early cinema itself. It could be seen, for example, that Fitzgerald's role - clarifying short excerpts of film with the aim to elucidate and inform - offers links with the lecturer used in early film exhibition. This was the lecturer who, shaped by industrial and social shifts in the development of film form, appeared around 1908. As Tom Gunning explains, this lecturer was endowed with educational connotations of cultural uplift and he played a different role than that of the earlier showmen. As an integral part of the cinema of attractions, the latter functioned 'much like a carnival barker' as he '*hyped* the films as extraordinary illusions and scientific marvels', heightening the effect of the spectacle on screen (Gunning, 1991, p.91). The BBC, of course, specifically wished to avoid the impression that they were 'hyping' or 'advertising' the films, and it is intriguing that McGivern invokes the 'showman' or 'barker' in his criticism. Given what he termed the 'comparative intimacy of people's homes', it was precisely these public and 'commercial' connotations which the BBC wished to avoid.

With regard to the excerpts themselves, the issue of the 'trailers' preoccupied the trade press. This was particularly so in the case of the technological limitations of television, what *Kine Weekly* termed 'the disability of being "boxed up" in black and white... and viewed by the light of the flickering fire'.[31] These concerns played an important role in the film industry's decision not to supply full-length features to television. As John Davis, Managing Director of the Rank Organisation, explained: 'If televised, feature films will lose much of their quality and definition and in fact give the impression that films are not good. Thus, we depreciate the quality of our own goods' (quoted in Macnab, 1993, p.201). It was clearly perceived that when screened on television, films promoted cinemagoing more generally. This traded upon a conception of filmgoing as an indiscriminate habit, existing as part of a weekly routine, although this was precisely a pattern which television itself was hastening into decline. As with the full-length film, however, the film industry were concerned that *Current Release* might in fact be bad publicity for their products and with current films, the stakes were much higher. While such concern was understandable (particularly at a time of declining admissions), it seems to be based on a rather simplistic view of the audience. They were likely well aware of the limitations of television and thus the technological superiority of the 'big screen'. Nevertheless, the film industry insisted that Fitzgerald emphasise when a film was in colour (something deliberately foregrounded by Susan's comments such as 'You won't forget to say it's a *colour* film, will you John?'), and this complemented their marketing strategy which attempted to differentiate film and television. As discussed in detail in Chapter Seven, the film industry's strategies in this respect reach their climax later in the decade with the increasing production of CinemaScope films and the 'X' certificate which, like colour, were intended to offer the public what television could not. However, this argument is partly problematised by the cinema programme given its active involvement in *promoting* these shifts on a number of different levels.

But in this earlier stage with *Current Release*, the film industry were concerned not only with what they perceived to be the technological limitations of television, but also about the content of the excerpts - which aspects of the film were shown in the preview. The BBC retained complete control here and the film interests sometimes only encountered the clips when *Current Release* went on air. The film companies were occasionally required to provide production facilities and, despite the misconception that the publicity was free, they paid the costs of preparing the excerpts. Companies later complained about the excessive nature of these charges (up to £130), but their payment is again indicative of their willingness to collaborate with television, and to experiment with its possibilities. Nevertheless, what the film industry termed 'the BBC's idea of trailer treatment'[32] often perplexed them. Following the first edition, *Today's Cinema* considered that the clips could 'have been chosen with a sharper eye to dramatic effect'[33] while *Kine Weekly* complained that *Lady Godiva Rides Again* was 'stripped of its humor' and that the sequence from *Encore* 'gave no hint of the romance of the central story'.[34] Perhaps most revealingly *Kine Weekly* later insisted that '*The Day The Earth Stood Still* missed its target. The publicity campaign is built around the robot, but the robot didn't appear in the clip'.[35] This raises the question as to the distinction between the cinema trailer and the television excerpt, an issue which Gerald Cock had broached in the earlier days of pre-war television. *Current Release* certainly advertised the films it screened, yet the way in which the press branded the series an example of pure 'plugging', or directly equated it with the film industry's *own* advertising, obscures the ways in which the television excerpt differed from the cinema trailer. While this may seem only a minor point, it speaks to the discourses and conceptions which accompanied the entry of film into the home.

'The BBC's Idea of "Trailer Treatment"': The Cinema Trailer and the TV Excerpt

An article published in *British Kinematography* in 1953 entitled 'The Production of Trailers', provides an excellent insight into this area from the trade's point of view. [36] Firstly, it is clear that there are certain similarities between the two forms. *Current Release* occasionally used a straight cut from a film but as with the cinema trailer, the excerpts were often constructed from several extracts edited together. According to the producer of *Current Release* this was often a necessity, for a long sequence could 'give away the plot of the film', something which he 'always tried to avoid'.[37] Restricted knowledge was also vital to the promotional function of the cinema trailer for as *British Kinematography* explained, 'You can lead up to your climax... but the cardinal sin is to give away the result... Leave [the audience] hanging by their finger tips' (p.99). This, however, is where the similarities end. An important distinction was the type of narrative constructed, what the trade termed 'the trailer story' (p.98). With regard to trailers produced for cinema exhibition, the shots were often intended to mislead, they were specifically understood to be out of context. As the article explained:

> For instance, in the case of a mystery story, any isolated shot which suggests mystery would be utilised, although in the feature there may be nothing mysterious about the shot at all. Shots such as these, out of context, can be... useful to give the trailer a build-up of atmosphere

(pp.98-9).

It is thus not surprising that the trade paper emphasised how 'The continuity of trailers scenes is not of great importance' (p.99). In terms of *Current Release*, the situation was different. As programme material, the film sequences had to be directly intelligible to the viewer, and McGivern also favoured clips that were 'self-evident' or self-explanatory as they 'minimised the tendency to "build up" or "sell" a picture unnecessarily'.[38] On certain occasions, however, the television excerpt also lacked continuity. As Fitzgerald explained when introducing the MGM musical *Skirts Ahoy* (Sidney Lanfield, 1952), 'You can see we haven't tried to sort out the plot for you - we just dipped around in the picture to provide you with some of the featured items'.[39] Both the viewers and cinema trade would sometimes complain that a particular sequence was 'quite meaningless'. Unlike the cinema trailer, they did not expect the shots to be out of context, they were expected to offer some form of coherent meaning in themselves. It was not simply a case of linking scenes arbitrarily, as *Picturegoer* emphasised when it published a review of *Current Release* entitled 'Trailers in Trouble':

> Television's new film feature, Current Release, still requires a lot more polish. The idea of running trailers is basically sound, but the handling of the chosen excerpts isn't all that happy, by a long way... It's obvious that the [BBC]... haven't got the measure of the film trailer technique in the same way as they've got it for fireside news-reels.[40]

While referring to the sequences as 'trailers', *Picturegoer* acknowledges that the television excerpt should differ - that it demands a particular 'technique'. The reference to the 'fireside' also suggests that this is inextricably linked to its domestic exhibition context, and connotations of intimacy and proximity.

An important difference between the two forms was what the 'trailer story' was intending to convey. *British Kinematography* explained how, in constructing the cinema trailer, careful consideration is given to the angle used to publicise the film:

One wonders whether the Stars are big enough; are the scenes in themselves good enough to sell the seats; how much padding do they need? Does the Director mean anything to the public and is he worth emphasizing? Is it a controversial subject, or should one make it so. Dare we sell it on sex, without making it nasty or running into censor trouble; or if it is a funny film, will the excerpts, divorced from their complete build-up, be funny enough?

(p.99)

As discussed, *Kine Weekly* considered that on *Current Release*, *Lady Godiva* 'was stripped of its humour' and felt that 'The quick collection of ten-second extracts were not ordinarily amusing'. The sequence from *Encore* was perceived to have 'missed its target because it gave no hint of the central romance of the story'.[41] *Kine Weekly* is emphasising the perceived sales angle as described by *British Kinematography* - in this case the genre of the film - indicative of the degree to which the trade were approaching the programme from a promotional point of view. *Current Release,* however, did not operate on the basis that it had particular advertising points to convey. The excerpt was simply intended to establish the plot or general themes which may, or may not, coincide with its publicity angle. Recall *Kine Weekly*'s complaint when discussing *The Day the Earth Stood Still* that the 'publicity campaign is built around the robot, but the robot didn't appear in the clip'.

Perhaps one of the most important differences between the television excerpt and the cinema trailer is what we might call *pace*, as recognised by *Picturegoer* when it commented on the 'technique' of the television clip. This was partly shaped by the exhibition contexts in which the material was viewed. While *British Kinematography* insisted that the industry 'no longer used the "super-colossal" adjectives' attributed to them, it acknowledged that 'Trailers are much larger than life and a little noisier too. We recognise the fact that they are wedged in-between the popcorn rattling, seat-tipping and ice-cream sales' (p.98). This was clearly very different to the domestic exhibition of television. Conceptions of the 'private "intimate" sphere of the home' (Jacobs, 2000, p.7) were often foregrounded not only in discourses on television aesthetics and viewing, but when it came to the broadcasting of feature films. Maurice Gorham explored this issue in *Television: Medium of the Future* (1950):

Unless television as a medium is to throw away its greatest advantage, its prime audience will always be a home audience, and a small group in the living-room will demand a different tempo and a different feeling in their television programmes from what they will welcome when they visit a theatre on their evening out... The stridency of musical openings to may films would blow viewers out of their armchairs, and it is hard to imagine the typical film trailer, all explosions and superlatives, raising anything but a laugh in the home'

(p.31).

In rejecting the pilot script for a cinema programme produced by the film industry, the BBC had perceived its address undesirable in view of the 'comparative intimacy of people's homes', and in combining a live direct address with the film excerpts, *Current Release* had to be careful to create a text considered appropriate for the domestic context in which it was viewed. Concurring with the view expressed by Gorham on the particular 'pace' and rhythm of television's mode of address, the BBC's film sequence manager explained that they 'should reduce the tempo of [*Current Release*] from that of the cinema to something more suitable to television'.[42] As such, it is worth noting here that the aesthetic relations between film and television are now often presented as the *reverse* of perceptions in this earlier period. Television is often seen to have a more rapid editing pace than film - leading to a 'corruption' of the cinematic aesthetic. Mark Crispin-Miller claims that now films are most likely to be seen on the small screen, 'Visual and spatial scale are down-sized, action is repetitiously foregrounded... pace and transitions are quicker... [and] music and montage ... more prevalent' (cited in Schatz, 1993, p.32). In the 1950s the aesthetic similarities and differences between film and television were being worked out, debated and, crucially, constructed.

Negotiating Film Criticism

As James Chapman and Christine Geraghty emphasise, films have always been 'produced, distributed and exhibited within a wider cultural context of criticism and debate to which filmmakers, critics, fans and academics all contribute' (2001: 2). I have emphasised the concept of 'film criticism' as in many ways central to the discursive debate and formation of these programmes, as well as their previous heritage in radio. But at the same time, it is in many ways difficult to explore them further within the context of 'film criticism' given the parameters in which they functioned. One of the reasons that this book is primarily concerned with the wider 'culture' of these programmes - the emphasis on sets, stars, premieres and so on - is precisely because this was their dominant textual focus. As a framework for the excerpts, the cinema programmes in the 1950s were to include very little negative criticism, presenting an overwhelmingly positive picture of the films it screened (and it is questionable whether they could in this sense be conceived as 'reviews' at all). Perhaps intending to give the impression of some measure of critical independence, Fitzgerald would sometimes refer to what he perceived to be a minor fault with a film. When previewing *Lady Godiva Rides Again* he suggested that 'The characters are all just slightly over done, but unmistakably human all the same...'[43] When it came to *The Day the Earth Stood Still*: 'A little more could have been made of this striking situation, but it's quite vividly done all the same...' We might compare this to a typical radio film review, such as that broadcast on the musical *Cover Girl* (Charles Vidor, 1944), starring Gene Kelly and Rita Hayworth. As the critic explained: 'when you try to recall the bits that you particularly enjoyed, nothing special will stand out in your memory. At least it didn't in mine'.[44] As these examples make clear, on television, any negative comments were always combined with a positive validation of the film, and the BBC's agreement to abide by the 'no criticism' clause indicates how keen they were to foster relations with the cinema, despite the possibility of attracting considerable criticism in return. This need not be interpreted, however, as indicating that the programme was simply a 'favour' to the film industry. It is clear that the BBC perceived that part of their public service responsibility was to help and support the British film industry, particularly when it was already perceived that television was likely to deplete its trade.

Yet Fitzgerald would also often incorporate small doses of education into his general commentary. In describing the plot of *Death of a Salesman* (Laslo Benedek, 1951) he saw an opportunity to elaborate on the complexities of the flashback, and when previewing *Home at Seven* (Ralph Richardson, 1952) he was keen to remark:

> This is taken from a stage play... and it's been made in a rather special way. [This]...
> is of particular interest to this programme because the production methods involved
> in the film... have been influenced by television technique. One could advance that
> argument very considerably, but I think it's probably enough to say... that the way the
> scenes are built up by the use of close and medium-close shots stems directly from
> television methods. Anyway, we'll see what you think.[45]

Such remarks gave the impression that there existed a much wider, 'specialist' knowledge on the aesthetic complexities of film and television, while simultaneously acknowledging that *Current Release* was not the place to explore this in detail.

More evident was the desire to place films in their wider social and cultural contexts in a way that has not been part of the cinema popular programme in subsequent years. An instructive example of this is the presentation of the Russian film, *The Fall of Berlin* (M. Chiaureli, 1949), released in Britain and America in 1952. One of few foreign-language films to appear on *Current Release*, it offered a Russian perspective on World War II, what *Variety* termed 'the Soviet Union's answer' to the many American and British depictions of war.[46] *Variety* also noted that 'With the increasing tension between the western world and the Soviet Union over Berlin, [this film]... has some contemporary significance'. *The Fall of Berlin* was previewed prior to its general release and as Fitzgerald explained, this was because it 'deals with a subject that touches us all very closely at the present time'. The BBC invited their Canadian correspondent, Matthew Halton, to appear on

Current Release and as Fitzgerald explained, 'We thought it would be of interest to our viewers to have your reactions to the picture'.[47] Halton began his narrative by relating his first-hand experience of the war: 'I saw Berlin immediately after its fall, when it was still burning....' *The Fall of Berlin* was constructed as the showpiece of the edition while the following contrast of the British comedy, *Curtain Up* (Ralph Smart, 1952), hardly received a mention. Fitzgerald simply suggested that 'There's sentiment, romance and a spot of drama too - to keep you going along between the laughs'.

In the case of *The Fall of Berlin*, the emphasis is on film's identity as a mediation or representation of historical events (and in this instance it is linked to the discourse of current affairs). Although outnumbered by the appearance of directors or stars, such items involved the use of an 'expert' interview, someone with often first-hand (and respected) knowledge of the subject in hand. This was not part of *Picture Parade* or *Film Fanfare*, nor indeed cinema programmes since. As discussed in Chapter Two, when certain critics in the 1970s complained about the 'benign', 'bland' and commercial nature of the film programme, they argued that it should be doing more to explore how films are shaped by the social context from which they emerge. The films should be presented, they argued, as not 'simply mental chewing gum, but as evidence of how the cinema encapsulates and encourages social change'.[48] While *Current Release* may fall short of these aspirations (which are arguably unreasonable for a cinema review programme screened at prime time and aimed at a mass audience), it nevertheless attempted to probe this area from an early stage - significantly when the *cinema* was still more of a mass medium. Four years later when *Picture Parade* and *Film Fanfare* were competing for an increasingly mass audience these possibilities were clearly limited, and the guests are confined to directors and stars. In 1952, however, these 'expert' interviews are important in understanding how the excerpts (and the general topic of cinema) were presented to the viewer as *television* material. From the BBC's point of view, this involved extending the boundaries of the film text to touch upon its wider cultural, social and historical contexts.

These items were nevertheless assimilated into the overall review format of the programme and were occasional, rather than regular features. They were greatly outnumbered by interviews with directors or stars, although when compared to later years, interviews were not a prominent aspect of *Current Release*. (Out of the thirty-two editions, fifteen included no interviews at all. In contrast, *Picture Parade* and *Film Fanfare* included up to four interviews in each edition). Nevertheless, it is already evident in *Current Release* that the cinema programme had the potential to increase the cultural visibility of the film director, and generate discussion of their practical, artistic and industrial role. The word *visibility* is key here given that, as mentioned in my discussion of radio programmes, coverage of the director's role had been a feature of the relations between cinema and broadcasting for some time. Despite this context, it is also difficult to determine the extent to which the concept of the director was part of popular cinemagoing culture: did filmgoers know the names of particular directors, or pursue their work? Not according to critic Stanley Reed in the same year that *Current Release* emerged: 'Most filmgoers know very little about directors. Instead they follow the "stars"', and he continued: 'A safer way to find the best films is to follow the directors' (1952, p.115). Addressing the reader directly, he enquired: 'How many directors do you know by name? One of the first steps toward more enjoyable filmgoing is to begin noticing the names on the "credit titles" at the beginning of a film' (p.115), and Reed cites a range of British and American directors whose work we might follow, ranging from David Lean, Anthony Asquith, Alexander Mackendrick, William Wyler to John Huston. Although not distinguished in this list, there is also the issue here of conceptualising the differences between British or American directors given that British cinema (traditionally perceived as in some way 'authorially impoverished') has historically found itself ill-served by 'auteur' theory (Hutchings, 2001, p.31).

When Fitzgerald introduced the films on *Current Release* he would sometimes mention the director, as well as the stars of a film, although he seldom elaborated on the significance of this. More interesting are the occasional interviews with directors, and an example here is David Lean. Before introducing *The Sound Barrier* (David Lean, 1952), Fitzgerald seemed to assume that Lean's name had a certain currency with the viewer, explaining that: 'I expect you know that before directing *The Sound Barrier* he was responsible for some of the most distinguished British films... [list of films] All of these films have one thing in common - a very arresting opening'.[49] Then addressing his question to Lean, Fitzgerald enquires, 'Do you have a special theory about that?' It is perhaps tempting to see this as prefiguring film studies' development of auteur theory with its insistence on the director as the locus of critical value and individual expression, and it is particularly interesting to note that (in the French journal *Cahiers du Cinéma* in the mid-1950s) this emerges at the same time as television's coverage of film expands. An investigation of these parallels could prove revealing, as could a wider consideration of how television has shaped the cultural circulation of the concept of the film director. In *Current Release*, however, rather than a unique, individual artist, Lean is perhaps primarily constructed as a skilled 'craftsman' and (particularly given that he discusses the adaptation of *Great Expectations* from novel to screen), as a 'translator' of a story into images. As Lean talks the audience through different scenes of his films and explains why particular aesthetic choices are made, it is clear that what television added to such coverage was the possibility of a visual demonstration. *Current Release* was keen to take advantage of this, and suffice it to suggest here that such items were again clearly intended to raise the viewers awareness of the aesthetic and artistic construction of film - perhaps subscribing to Stanley Reed's view that this would make for more 'enjoyable filmgoing' for the audience (1952, p.115).

In terms of the stars, the producer complained that the film industry was uncooperative in facilitating appearances, and that the stars themselves often failed to respond to his requests. After five editions of the programme he (justifiably) felt that all of the interviews so far had been 'with persons of rather secondary importance'[50] - largely less prominent British actors and actresses (and often up-and-coming ones at that). The BBC were informed that MGM British in particular had received instructions from New York that stars were not to appear on live television, although they were considering the possibility of recording interviews on film.[51] The other American companies appeared to take a similar approach and in response, the producer reported that a London representative of the Motion Picture Producers Association (MPPA) was to visit New York to 'acquaint... the Hollywood industry with the nature of *Current Release* and the advantages ... of co-operating in the matter of star appearances'.[52] With the costs being met by the film industry, the producer was also to attend for 'the purpose of securing at least 8 pre-filmed interviews of [sic] top-ranking stars - together with one or two short magazine items on Hollywood itself'.[53] It is notable here that 'top-ranking' stars are 'naturally' conceived as Hollywood performers, and it is certainly true that, aside from the previews of the films themselves, the cinematic 'world' constructed by *Current Release* was primarily comprised of British, rather than Hollywood cinema, the increased centrality of which had to wait until the more ambitious *Picture Parade* and *Film Fanfare* later in the decade. (Actors and actresses appearing on *Current Release* included Vera Ralston, Evelyn Keyes, Claire Bloom, George Cole, Eva Bergh, Ronald Shiner and a very young Joan Collins. Aside from a filmed interview with Bing Crosby and Bob Hope, perhaps the most widely known name was James Robertson Justice, although he was arguably regarded as more of a character actor than a 'star'). Hilmes explains how the Hollywood majors had expressed concern over broadcast appearances by film stars since their earlier development on radio. Exhibitors in particular expressed concerns over stars giving performances for 'free' (in so far as the listener is not paying a fee at the box office), as well as the danger of people staying in to listen to stars rather than going out to the cinema (1990, p.57). In particular, with an implicit perception of the broadcast emphasis on intimacy and familiarity as antithetical to distance, 'aura' and glamour, concerns were also expressed over film stars effectively 'using up' their box office appeal and exhibitors raised objections that their public image may be 'damaged' by entering into 'ill-considered or simply ill-performed stints on air' (Hilmes, 1990, p.59). As we will see, the worry that a star may jeopardise their public image by appearing live and 'unscripted' emerged as a particularly acute concern around the television interview. It was again the Hollywood companies that were the focus here, perhaps due to the fact that the star had always been more economically and culturally central to Hollywood cinema on a number of different levels, and the degree of control the studios had exerted over their contracted players. Television also emerged at a time when, with the reorganisation of the studio system (see Balio, 1990), studios no longer held stars under the kind of long-term contracts of the past - hence resulting in less control over stars' public appearance and their forays into other media forms.

At both a conceptual and historical level, the entire concept of film stars appearing on television requires careful analysis. As the 1950s progressed, television played an increasingly important role in the cultural construction of the film star image. Given that, in the British context, it was not until later in the decade that these appearances became institutionalised, it is more fruitful to consider the issue in relation to *Picture Parade* and *Film Fanfare* (Chapter Five). Nevertheless, the producer of *Current Release* was evidently keen to compensate for the relative absence of important stars as in a special 'Christmas Party' edition, no less than thirty-two guests appeared. While including many of the lesser-known names who had already appeared, the list also promised some more prominent British stars such as Dirk Bogarde, Margaret Lockwood, Richard Todd, Jack Hawkins, Anthony Steel and Trevor Howard. With Lionel Gamlin and Eamonn Andrews conducting the interviews and Fitzgerald presenting the films, the parade of personalities made entrances and exits throughout the programme, concluding in an item where all the 'guests' participated in a song with the British singer and actress Petula Clark.

The programme received a special billing in the *Radio Times* which explained 'You are invited to meet the following guests...', then listed the entire thirty-two names scheduled to appear.[54] This constructed the programme as a personal encounter between viewer and stars, a social gathering which the audience was invited to attend. The BBC clearly understood the event to be a glamorous party occasion as an internal memo stressed that it would be 'a super edition of *Current Release*.... very much evening dress, especially the girls'.[55] Yet in conjunction with complaints from the film industry which insisted that the programme was poorly organised and 'amateur', Maurice Wiggin of the *Sunday Times* was particularly unimpressed and described it as a 'mockery' and an 'abysmally unfestive party, conducted on the level of a "fan" magazine of the trashiest sort. It must (I hope) have made many easy-going viewers pause'.[56] Wiggin's response suggests that, from his point of view at least, stardom and showbusiness were not entirely appropriate for television, particularly a service run by the BBC. Such 'trash' was to be confined to other media such as the fan magazine and as a television critic, this is a perspective which Wiggin views with disdain. While the BBC at this stage clearly experienced difficulties in securing the interviews, it was perhaps not only practical problems which explain the relative absence of stars. As Wiggin's complaints suggest, there were also conceptions as to what was appropriate television material. In 1952 this embraced the live outside broadcast, the newsreel, and television's ability to enlighten and inform. It did not yet encompass 'showbusiness'. The BBC changed their attitude toward this later in the decade when faced with the increasing competition from ITV.

'Added Cinema Pleasure': The Reception of *Current Release*

While the film trade offered their own form of reception where *Current Release* was concerned, the reactions of the general audience suggest a different perspective. What does the consumption of the programme suggest about early viewing habits and television's entry into the British home? What does the reception of *Current Release* indicate about how it was used by its audience in the context of shifting trends in social leisure? How do responses to the programme reflect back upon the tensions which surrounded the cinema programme between the film industry and the BBC?

Historicising 'Tele-viewing'

As with all historical audiences, there are conceptual and methodological difficulties involved in researching television viewing and viewers in this early period. Unlike contemporary research into contemporary TV audiences, the most available information tends to be in the form of statistics and large-scale quantitative data used to demonstrate the rapid diffusion of television into the home, often neglecting the 'experience and contexts of television *viewing*' (O'Sullivan, 1991, p.159). Rather than using available traces of audience responses to the medium or the television programmes they viewed, one approach has been to examine the range of discourses which circulated around the entry of television into the home, whether in terms of its status as a domestic object, new technology and/or as a site of social concern, or site of gendered address (Spigel, 1992, 1997, 2001, Parks, 2002). This is work which has again predominantly, although not exclusively, originated from the American context (see Oswell, 1999). In contrast, Tim O'Sullivan's 'Television Memories and Cultures of Viewing, 1950-65' (1991) adopts an empirical approach by asking people to recall their early experiences of television in Britain - although this necessarily involves engaging with the issue of memory, and all the methodological issues this may pose.

Here, I combine a focus on the discursive construction of early television viewing with the analysis of responses to *Current Release* - as collected from certain surveys at the time. From the perspective of audience viewing patterns, the cinema programme would seem to be situated within contradictory and transitional trends in social leisure. As discussed in Chapter One, the post-war period is understood to have seen an increasing shift toward more home-orientated leisure or privatisation, contributing to the cinema's decline. While television was a significant influence in the growing investment in the home as a site of leisure, it was only one of a number of contributing factors which included demographic shifts and rising standards of living. Graham Allan and Allan Crow caution that 'We need to be... sceptical of the stereotypical model which suggests that at some time past, social life was more communal and less focused on the household... than it currently is' (1991, p.19). There had of course been a good deal of domestic entertainment before the 1950s, not least of all in the form of radio. While public pursuits such as the cinema and the public house may have suffered from the shift toward home-orientated consumption, the 1950s in fact saw a proliferation of leisure opportunities which (with the gradual expansion in car ownership) took people away from the home (see Hopkins, 1963, p.430). In any case, the cinema programme complicates a dichotomy between 'public' and 'private', since it simultaneously offers an insight into changing attitudes toward cinemagoing and the increasing impact of television on daily life. While this might be conceived primarily within the framework of the further domestication of film culture this is, as discussed, very much a transitional period for the cinema. Before exploring these relations, it is important to consider television's entry into the British home, and the type of context in which *Current Release* was consumed. Discourses on early viewing are crucial in understanding the responses to the programme, and can illuminate more general viewing habits in these early years.

Advertisements, the popular press, magazines and books, as well as the institutional publications of the BBC, all provide forms of commentary on the increasing spread of television and its entry into the British home. As Spigel explains, these offer discourses which 'do not reflect directly the public's response to television [but]... they do begin to reveal the intertextual context through which people.... might have made sense of television and its place in everyday life' (1992, p.4). The expansion of the medium was more rapid in America than Britain, and as acknowledged, in the immediate post-war years the ownership of a television set was restricted to the more affluent home. When *Current Release* began in 1952, American women's magazines referred to television as a 'family pet' or 'family member' (Spigel, 1997, p.82), while the British were still being advised as to how to install the set when it first arrived.[57] Similar to Spigel's work on the American context, David Oswell describes how discussion in popular magazines advised British readers on the appropriate placement of the set and its assimilation into the geography of domestic space (1999, p.68). Television was an unfamiliar and almost 'alien' technology which was yet to be assimilated into the domestic landscape of home, and this discussion was often enmeshed within concern regarding television's impact on domestic space and the construction of what Lynn Spigel terms the 'domestic gaze' - how to organise an economy of vision in the home (1992). As well as being instructed as to how to situate the set, early 'televiewers', as they were called, were also informed how to view. As the *BBC Yearbook* in 1946 insisted:

The television set demands your attention; you cannot enjoy it from the next room.

You must sit facing the set, with the lights down or shaded, and if you are a normal viewer you will find yourself very reluctant to be disturbed

(Gorham, 1946, p.20).

The use of the term 'normal viewer' suggests that there quickly developed prescriptive and *desirable* ways to use the medium. Jan Bussell's *The Art of Television* (1952) articulated the importance of this more forcefully:

The habit of commenting and talking in a programme, so easily formed, is a very bad one... It results in points being missed and the loss of atmosphere and mood, besides being discourteous... It is no use approaching the set with an 'amuse me if you can' attitude. Make sure you are the right distance from the screen- not too near and not too far... most screens are bright enough to be viewed in broad daylight [but] it is far better to darken the room [to] shut out local surroundings and thus concentrate to the exclusion of all else

(Bussell, 1952, p.117).

Although problematic to imagine a dramatic 'retreat' to the home at this time, in Bussell's description, viewing is indeed imagined as intensely and pleasurably secluded. This is arguably complicated, however, by the phenomenon of 'guest' viewing. As not everyone possessed a set at this time there was good deal of viewing at a friend or relative's house until well into the 1950s, and this suggests that the technology was not fully domesticated, nor was the experience of viewing yet private or exclusive (Caughie, 1991a, p.27). In contrast to the unease surrounding television as an 'intruder' and technological eyesore, the comments above suggested ways in which to gain particular pleasure from viewing in the home. This supports Spigel's argument that television ushered in a theatricalisation of domestic space in which 'The home was figured as a kind of "ideal theater" where visual pleasures achieved new heights' (Spigel, 1992, p.4). Some of O'Sullivan's interviewees recall the fascination of having the broadcast picture 'in your own home', like 'having a private cinema or newsreel' (1991, p.166). The idea of a group audience sat in darkness, absorbed by the image on screen, clearly evokes connotations of cinematic spectatorship. Indeed, in relation to this early period, it would be difficult to apply John Ellis' famous conception of television viewing as being organised around an economy of the 'glance' (in contrast to the economy of the 'gaze' in cinematic spectatorship) (Ellis, 1992, p.163). Modes of consumption, and their discursive construction, are historical. In fact, a term initially suggested to describe the TV viewer was 'gazer' (Briggs, 1965, p.114). This image of a theatricalised, cinematic, yet profoundly *domestic* space offers the possibility of a particularly intriguing context for the consumption of the cinema programme.

This organisation of visual pleasure in the home was also seen to have a potentially disruptive effect on domestic space and routines. A critic of the time noted that 'All new inventions create panics' (Brown, 1952, p.17), and the advent of television was marked by debates regarding its impact on the home, the family and daily life (Spigel, 1992, Oswell, 1999). The BBC urged the public to view selectively and discriminately and as John Caughie notes, it was only the BBC's monopoly as a public service broadcaster - before the competition from commercial television - that enabled them to articulate this paternalistic 'anxiety about the family and the home' (1991a, p.27). On one level, television was promoted as contributing to the hope for a return to domestic values and family cohesion after the disruptions of war, but it also gave concern for anxiety, threatening the everyday rhythms of family life that had come to assimilate the apparently less commanding distractions of radio. While Brown insisted that television was 'stabilising the Englishman's home' (1952, p.17), a critic in *BBC Quarterly* believed that viewing was not a family activity at all:

> I am not impressed when I hear that television keeps the family together... they are no more together than a number of monologues in the same room constitutes a conversation. It is not the same as being a team, each doing some bit in a common interest... game or practical necessity... It is the ... enjoyment of several private experiences

(Demant, 1954, p.139).

A range of opinions, then, circulated around the relations between family life, leisure, and television, and these are also evident within, and further articulated, in the responses to *Current Release*.

In considering the audience, I am interested in three broad areas: the programme's role within the economy of early viewing and as part of this, its effect on viewers' cinemagoing. This then enables a consideration of how *Current Release* was perhaps used by its audience (or at least those surveyed) and how this reflects upon its relations with both advertising and public service. In the institutional debates which surrounded the programme these principles were rather inflexibly opposed, yet in order to fulfil either function, *Current Release* must ultimately meet with its audience.

Exploring the reception of the programme has involved the use of three key sources: the BBC's own research on audience responses to the programme (daily surveys of audience reactions to television programmes were established in 1950), an extensive survey on the programme conducted by the cinema trade, and a wider survey published in *Picturegoer* which investigated the impact of television on cinemagoing the same year *Current Release* began. From the early 1950s,

Picturegoer published letters which suggested how the cinema might compete or co-operate with its small-screen rival. By 1952 these comments appear more frequently and in turn, shape the content of the magazine. In February 1952 *Picturegoer* announced its 'Films and TV' survey which enquired, 'What effect - if any - is television having upon the film habits and appetites of British picturegoers?'[58] The forum was partly prompted by a letter which explained:

> As a picturegoer and TV viewer I feel that the cinema will have to look to its laurels. I used to attend the cinema twice or three times weekly. Now, sometimes a week passes without my seeing a film. I stood in the rain to see [Desert Fox The Story of] Rommel [Henry Hathaway, 1951] as I knew it was a spectacle that TV could not reproduce, but if I'm to endure such discomfort in the future, the films will have to be good. If the cinema's counter-attraction is not strong enough, I feel that to sit by the fire and turn a knob for an evening's entertainment may well become a national habit... I wonder if other readers agree with me.[59]

While the survey was limited in its scope, the responses can be used to complement the trade report on *Current Release*, and to build up a wider picture of television's impact on filmgoing. The film industry survey displayed the rather ambivalent title '*Current Release*: Good or Bad?', and was published as an eight-page supplement to *Today's Cinema* one year after the programme began. *Today's Cinema* explained how *Current Release* had been a productive 'talking point' for the trade, particularly in terms of its possible effect on cinema admissions.[60] *Today's Cinema* claimed that the survey was prompted by the desire to explore this issue, and it was conducting across several different regions,[61] providing over eighty responses to the programme.

There are certain methodological problems involved in using these sources as historical evidence. The responses are often decontextualised, lacking indication as to the gender, age or wider personal background of the respondent. The BBC and the film industry were also doing different kinds of audience research, with different investments and aims. The BBC aimed to discover if the programme was generally enjoyed by viewers, while the *Picturegoer* and trade surveys were interested in its effect on cinemagoing, with *Today's Cinema* in particular concerned with its value as a promotional tool. As the responses will suggest, these areas are not mutually exclusive, yet the differing objectives shaped the questions asked and thus the nature of the responses received.

While the cinema programme was not simply a substitute for the feature film, nor a bargaining counter with which films might be secured, it is nevertheless the case that the relative absence of feature films on television shaped the *reception* (and promotion) of *Current Release*. While providing a different experience to the full-length feature, *Current Release* was nevertheless promoted as a place to see *film*.

Films were not yet televised on a regular basis and those that were comprised part of the BBC's somewhat limited library. Feature films were sometimes scheduled during the day (in 1952, for example, this included British comedies from the 1930s starring George Formby or Gracie Fields), but evening screenings were infrequent. Those that did appear were often also regarded as excessively 'old'. Barbara Klinger (1998a) describes how 'Once a physically remote, transitory...medium, [film] has attained the solidity and semi-permanent status of a household object', and her reference to films as 'transitory' is important. The televising of films emphasised their extended commercial life on an unprecedented scale and as *Sight and Sound* acknowledged in 1957, 'The standard joke about films on television is their age' (Crow, 1957, p.62). The critic explained how 'The film industry [had] spent years indoctrinating the public with the belief that an old film is as dull as yesterday's newspaper and that only the latest release is worth seeing' (p.61). As a marketing strategy, the film industry promoted the cinematic product as ephemeral and transient. While the screening of 'old' films at the National Film Theatre usually attracted a limited audience (Crow, 1957, p.61), reissues were a regular part of cinema programmes at this time. However, the average age of reissues was approximately 4 to 5 years old, whereas for the televised films (excluding those in a foreign language), it was usually 8 to 12. This appeared to make a substantial difference, and the age of the films was often emphasised by their poor presentation on television.

While television was associated with screening very 'old' feature films, features so 'ancient' they were something of a joke, the cinema programme is evidence of the fact that such films co-existed alongside coverage of contemporary film culture. *Current Release* was not simply a place to see 'film' on television when full-length features were in short supply; it was an opportunity to see *new* film, products not yet glimpsed on the cinema screen. At the beginning of each edition, Fitzgerald lists each of the films being screened as if to 'hook' the viewer from an early stage. That the clips were perhaps the prime attraction was also suggested by the presentation of the programme in the television schedules, the expectations it sought to build, and the type of engagement it promised to the viewer. The *Radio Times* billing for *Current Release* emphasised how 'viewers will see clips from...', and listed the films which would appear. This often painstaking process remained the conventional way of listing cinema programmes throughout the 1950s, even when the stars became much more central to its culture. Responses from the audience also suggest how the cinema programme provided an important experience of film on television in the 1950s. Several viewers, for example, found it frustrating that more of each film was not screened. One viewer explained that while he enjoyed *Current Release*, 'it often aggravated [him] in the same way that the old-fashioned serial story did, by cutting off just when your interest had been aroused' (p.2),[62] and BBC Viewer Research revealed similar reactions. This not only emphasises how *Current Release* was perceived as a place to see 'film' at this time, but also indicates that a television programme comprised of excerpts was a new concept to viewers. Film programmes on radio often used excerpts, but the visual nature of television, and thus *Current Release*, engaged the audience in a new way. In general, however, available evidence on both the consumption and promotion of *Current Release* suggests that the wider absence of feature films on television played a significant role in shaping its reception.

In understanding the popularity of the programme it is also important to stress that in 1952-53, there was only a limited amount of light entertainment on television. *Current Release* was initially screened on Thursdays on a fortnightly basis after the repeat of the Sunday play. By July 1952 it had proved to be so popular that it was moved to Wednesday, described by the cinema trade as 'a key viewing evening' in the television week. It was screened at 9:30-10:15pm and while this was its regular slot, there was little consistency in the surrounding programmes. (One week it was *Pleasures and Painting*, 'a story of great artists' and the next week *A Fair Price?*, 'a report on the marketing of fruit and vegetables'. On one occasion Wednesday night viewing also brought the pleasures of *Happy Hoedown*, 'a programme of square dancing for beginners'). Different forms of drama occupied the television schedules at this time, and would continue to do so with the advent of ITV. In 1952 drama could be scheduled up to 4-5 nights per week, the majority of plays being adapted from the West End theatre circuit, the classics, or the bestseller lists (Caughie, 1991a, p.27). A letter to *Picturegoer* claimed that 'Television is little more than a photographed stage-play'[63] and while viewer research indicated that plays could be extremely popular, this highlights how *Current Release* was different from other television fare. Commenting on the popularity of *Current Release* a Viewer Research report noted that 'It achieved a reaction index of 70, a very good figure by any standards, though no television programme on the same lines has been reported on before'.[64] Viewer Research Bulletins suggested that *Current Release* was watched by between 50-70% of the television public. At a meeting between the film industry and the BBC the producer reported that while its audience had initially been around two million per episode, 'latterly the viewing audience was probably up to five million'.[65] Programmes such as the well-known magazine feature *Picture Page* were not as popular as *Current Release*,[66] but a feature which stood out at this stage was *What's My Line?* (BBC 1951-62, 1973-74) one of the first panel games on British television. It scored very highly with viewers and the BBC noted that 'It had established itself as a favourite'. As one respondent in the survey explained, *Current Release* was his 'second favourite' programme as *What's My Line?* could not be beaten (p.3).

Despite such responses, viewing routines were not necessarily organised around particular programmes at this time. This was partly due to the BBC's scheduling policy. It was not until the advent of commercial television that viewers were encouraged to form habits on a weekly basis and to a certain extent, the BBC had previously resisted this regularity. They did not want television to be used habitually or casually, and regular scheduling was seen as a way of encouraging this. While there were certainly regular programmes, these were as likely to be fortnightly or monthly features, as they were weekly. There were also many short series and isolated programmes designed to 'keep viewers on their toes' (Sendall, 1982, p.324). Indicative of the BBC's irregular scheduling policy, one correspondent in *Picturegoer* explained that she 'wouldn't stay at home to watch a pot-luck programme on television'[67] while (perhaps also referring to frequent technical breakdowns) another described television as 'unreliable' - responses which are hardly suggestive of established viewing routines.

In 1952 BBC Viewer Research suggested that the average viewer watched one hour of television per night (Silvey, 1977, p.164), but evidence indicates that many watched the evening programme from beginning to end (8-10:30pm). In the *Current Release* survey, a few respondents made such comments as 'I don't go for it specially, but I watch anything that interests me', or 'I see it through, along with the other features on offer' (p.5). It is interesting, however, that for the majority of those surveyed, it was *Current Release* was something of a highlight in television's weekly flow and an early example of organising preferences around programme specificity. To simply take one example, a viewer from Sheffield explained how he 'look[ed] forward very much to the programme. It is something off the usual TV-beaten track' (p.7). A Birmingham cinema manager reaffirmed this view when he acknowledged that, 'According to the public, *Current Release* is one of the most interesting things from an entertainment point of view that the TV people put on' (p.6).

'Stay-at-home Picturegoing?': TV Viewing, Cinemagoing and *Current Release*

Rather than simply being 'pretty poor value for the viewer', or 'a perpetual reminder of what TV might have but never gets' - that is, feature length films,[68] *Current Release* was in fact one of the most popular programmes on British television at this time. Yet it is crucial that this very popularity also gave the film industry cause for concern, reflecting on the programme's contradictory status within the context of prevailing shifts in the consumption of screen entertainment. Hilmes' work on the relations between Hollywood and broadcasting emphasises how, right back to the mass dissemination of radio, *exhibitors* in particular were concerned about the effect of broadcast coverage on cinema admissions (1990, p.56), perceived not only as a promotional tool which may enhance box office receipts, but also as a form of *competition* in attracting an audience for 'film'. Hence, their request that radio should curb star appearances during theatre hours (Ibid). It has always been the exhibitors which are most under threat from the advent of new sites for the distribution and exhibition of film given that their fortunes are inextricably wedded to the public venue of the cinema (Hilmes, 1990, p.42). Television was of course a key form of competition here, and while focused predominantly on the screening of full-length feature films, these concerns did not exclude the cinema programme. Reports in the trade press suggest that, in the US, a similar situation occurred with the increasing proliferation of television's coverage of film (particularly given its larger number of channels),[69] and it was a central feature of the negotiations between the film industry and the broadcasters in the British context. As paradoxical as it may sound today, from the beginning of *Current Release* the film industry had debated as to whether it may 'defeat its own object', that is, keep people *away* from the cinema on the night it was screened. This marked the start of a long debate over the cinema programme which continued throughout the 1950s. The film industry were essentially concerned about people's changing leisure habits and their fear was based on a rather homogeneous conception of the audience: they were perceived to be either watching television or at the cinema, although the trade's own perception here was contradictory as, particularly later in the decade, they would acknowledge that it was in fact a proliferation of leisure opportunities which were competing for patron's time. In the case of broadcasting, Ien Ang (1991) has famously argued that audience research is an attempt to exercise a discursive control over an 'unknown' and 'essentially unknowable' audience. While Ang was primarily referring to the use of ratings here, her point about audience research offering an illusion of mastery over what (in reality) is an elusive and unpredictable concept is relevant here. At a time of great uncertainty and change, the trade's audience survey can perhaps be understood within this context, but the film industry's attempt to exercise a 'control' over the audience took on more literal dimensions as the decade progressed. By 1956 the coverage of film on television had expanded considerably, and the film industry began to stipulate *when* the programmes could be screened. To a certain extent, this was an attempt to intervene in the trend of people's changing leisure habits, particularly the shift from the cinema toward the home screen. The film industry is still conceiving the cinema as essentially a mass medium at this point - and battling against the transition away from this - a situation which points to the cinema programme's contradictory role within this shift. While it may be perceived as a rather excessive action, especially as these programmes were promoting film, the film industry was anxious that they should not conflict with periods when patrons might otherwise attend the cinema, and the *Current Release* survey was the first intervention in this field.

One of the questions put to patrons was thus whether they would stay at home specifically to view the programme, and this suggests how the survey attempted to investigate the extent to which people's lives were becoming more 'home-based', as one exhibitor described it (p.5). As difficult and elusive as the subject of privatisation may be to explore, the responses to the survey seem to touch on this at certain points. Many respondents said that they did stay in to watch *Current Release* and as one explained, 'I would stay at home to see it as it gives me added cinema pleasure' (p.2). Notably, several who expressed such sentiments were women. Other female viewers suggested that *Current Release* did not conflict with their filmgoing as they often went 'first house', as one woman explained, if attending an early performance at the cinema, she often 'hurried home to ensure that she didn't miss *Current Release*'. According to this viewer, there was 'no difficulty in being at home between 9 and 10 [pm] to view it. [She] got quite a kick and thrill out of it' (p.5). The fact that women tended to go 'first house' which then made them available to view *Current Release*, was also noted in the trade summary of the survey (p.1). It would seem that some of the respondents used the term 'first house' to refer to daytime cinemagoing, while others used it to describe the first performance of the evening. Women had generally represented a substantial proportion of the daytime cinema audience, although such an opportunity was clearly dependent upon other factors, such as employment and childcare. Perhaps the increasing number of women in part-time work contributed to the decline in women's daytime cinemagoing by the late 1950s. According to Harper and Porter's study of 1950s British cinema audience, this traditionally represented what might be termed 'indiscriminate' filmgoing - when visits to the cinema are organised less around a specific film than they are part of a weekly routine. Yet they explain how 'housewives gradually lost the indiscriminate habit as the decade progressed. By 1957 over half [of those questioned] never went at all' (1999, p.68).[70]

On the basis of the survey, *Today's Cinema* perceived that *Current Release* appealed more to women in general (p.1). Several male respondents noted that their wives were 'more interested' in the programme than them and as one explained, 'It's because of that that I have come to think of *Current Release* as more of a woman's programme than a man's - mainly because... women have more time' (p.3). This observation is of course highly debatable, particularly given the work emphasising the extent to which early television viewing for women was often figured as a source of 'potential domestic guilt' in view of their household priorities and the patriarchal division of labour within the home (O'Sullivan, 1991, p.177) (Spigel, 1992, 1997). Based on the evidence here, it is not possible to pursue the question of whether *Current Release* really *did* attract a larger proportion of female viewers. Within the limits of the survey it could be suggested that if *Current Release* was perceived to be more of a 'women's programme', then this conception was perhaps shaped by the appeal of the presenter, John Fitzgerald. When McGivern expressed concern about the suitability of Fitzgerald the producer reassured him that the presenter's 'brand of easy charm [was] right for the programme at this stage'.[71] The term 'charm' has connotations of a sexual attraction and Fitzgerald certainly seemed to be popular with the female viewers. As one explained, 'I like the programme very much indeed, and I like John Fitzgerald who handles it. I think perhaps my liking for him... has something to do with my liking for the programme' (p.4), and other women expressed similar sentiments. It was not simply the films, therefore, which could act as a key attraction in the programme. Fitzgerald's physical appearance, demeanour and indeed 'TV personality' are also factors to consider here.

In terms of the programme's links with cinemagoing, a cinema manager from Liverpool suggested that *Current Release* would entice people to the cinema where they could see 'full pictures in an atmosphere more suitable for entertainment' (p.2). While a cinema manager has a vested interest in making such a claim, his construction of the comparison is revealing of contemporary attitudes. The suggestion that the home is not suitable for viewing screen entertainment would today be seen as strange - indicative of the shifts in leisure and technology which have occurred. A similar conception was also expressed in the *Picturegoer* survey when a reader emphatically stated that his 'idea of entertainment [was]... an evening out, not an evening in an armchair'.[72] Perhaps drawing on similar trends in cultural experience another explained, 'I could not look forward to an evening's entertainment on TV as I do to an evening at the cinema'.[73] Particularly in the first response, nightly entertainment is seen as incompatible with the domestic sphere. With the production of fewer, more costly films (eventually leading up to the 'blockbuster' trend), this is a view which complemented the film industry's marketing strategy from the 1950s onwards - promoting cinemagoing as a 'special event' in contrast to the routine availability of domestic screen entertainment. Innovations in technology and exhibition practices such as widescreen, colour, lavish epics or confectionery, were all designed, as Geoffrey Macnab describes, 'to emphasise the "specialness" of cinemagoing experience as a stark contrast to the banality of staying at home and watching the "box"' (1993, p.211). The survey makes clear how the notion of selective and 'event' filmgoing was already evident in the early 1950s as opposed to the possibly more habitual and routine approach of earlier years. Yet the responses which construct the domestic sphere as an inherently unsuitable space in which to consume film - and indeed screen entertainment in general - do not necessarily cohere with the perspective that Macnab describes. Articulated at the start of the 1950s, the responses reflect less a considered comparison between filmgoing and home viewing, than a more general ambivalence regarding *domestic* screen entertainment - the theatricalisation of the home - and the fact that it was initially strange to view moving images within this space. Macnab's description of the film industry's marketing strategy as designed to construct the 'specialness' of filmgoing in contrast to the 'banality of staying at home and watching the "box"' (Ibid) suggests a great familiarity (or even a boredom) with television. It is clear that in many cases, this familiarity had not yet taken shape, and the responses precisely articulate the perceived *strangeness* of the new technology, and its location in the domestic space of the home. While it is important not to forget that there had been of traditional domestic entertainment before television, in the above responses, home-centred leisure is constructed as almost an 'alien' concept. (To make a comparison with today, 'an evening in an armchair' would be less unfamiliar than conventional.)

In responding to the film industry's question regarding *Current Release*, some viewers expressed such sentiments as 'I don't stay at home specially to hear [sic] any TV programme' (p.5) or 'Not on your life would I stay at home to see it' (p.5). One viewer from Leicester provides an interesting response in this respect and explained: 'My wife and I enjoy [*Current Release*] very much and look forward to seeing it. We do not stay at home to see [it]; we carry on with our nightly entertainments as usual, and are not influenced by any TV programme' (p.4). These responses suggest that people were resistant to the question posed by the survey. Rather than simply indicating possible attitudes toward domestic entertainment, the audience responses were also influenced by the concerns surrounding television and its impact on the British home. While discourses on the 'effects' of television were more prevalent in the late 1950s (with respect to the child viewer for example) (see Oswell, 1999), television's impact on familial and social interaction was a topic of debate from an early stage. It was perhaps not surprising, therefore, that respondents were not keen to admit that they were influenced by the new technology, and their discussion of *Current Release* was informed by this resistance. That the replies were partly produced by a defensive reaction was suggested by the fact that those who claimed they were not 'influenced' by the programme (whether with regard to staying at home to see it or in other ways) would then happily explain how it influenced their choice of films. This was of course precisely what the film industry was seeking to explore.

One of the concerns surrounding television was that it was an unsociable activity, and this could relate to its general implications for social interaction or its more immediate impact on the family sphere. Letters in *Picturegoer* and the *Radio Times* (and later the *TV Times* when it began publication in 1955) debated this issue. A reader in *Picturegoer*, for example, explained how he 'had often gone to the cinema to get away from television when other members of my family have wanted it on',[74] and although from a slightly later period, another picturegoer described television as disrupting communal interaction and complained: 'I arrived from abroad to meet old friends and spend many companionable evenings. But I had to sit in gloomy dark and silence and watch TV. Now I'm giving "friendly" visits a miss - instead I'm going to the cinema'.[75] While shaped by the fact they are written to a film magazine and thus feel a duty to express their loyalty to the cinema, many letters suggest that the cinema could be constructed as a place for social interaction. Although this was certainly not new, it is explicitly foregrounded by these readers when compared to the experience of viewing television. In fact, in the early 1950s, letters to *Picturegoer* stress the communal participation of cinemagoing as compared to the more privatised and isolated experience of televiewing. As one reader insists, 'However good, TV can never compete with the atmosphere of a crowded cinema, where thrills are heightened and laughs shared with other picturegoers'.[76] Another explained that 'Our friend's description of new films and seeing our old favourites on TV made us long for the cinema atmosphere'.[77] Again implicit here is the notion that the home is not the most suitable arena for screen entertainment, and nor does it offer the appropriate atmosphere in which to view the feature film. In subsequent years this attitude appeared to shift as the material and social experience of the cinema underwent change. By the mid-1950s it was clear that the cinema, as Geraghty explains, 'was stuck with wartime facilities and was not keeping up with the publicised expansion of consumer goods and domestic improvements in the home' (2000a, p.7). There was also an increasing dominance of young people (or 'teenagers') in the audience. These factors worked to change the experience of the cinema as a social space and by the late 1950s, *Picturegoer* was littered with letters which compared the 'drafty', 'uncomfortable' and 'noisy' cinema with the more private tranquillity of viewing in the home (given that the magazine had increasingly pursued a youth audience by this time, these letters were presumably written by *Picturegoer*'s remaining older readership). In the early 1950s, however, the communal experience of cinemagoing was often valued as positive, although there were still those willing to forego such social pleasures (and perhaps accept the aesthetic inferiority of television), in order to enjoy the domestic comfort of home: 'Television has a long way to go before being first class, but oh, to sit by the fire, turn a knob and relax! What more can we want?'[78]

Clearly, the film industry's fears about people staying at home to view the programme suggest the ways in which television was beginning to impact upon cinemagoing patterns. The responses also indicate how, existing as a unique link between viewing *and* filmgoing, the cinema programme could function within this shift. Although the fall in admissions was not as significant as later in the decade, many respondents suggested that since owning a television set their filmgoing had declined and that *Current Release* thus provided their 'cinematic experience' in the shift toward a television-centred domestic leisure culture. Several commented that it was young children, or other commitments, which prevented them from attending on a regular basis. As a mother from Plymouth explained, 'I like [*Current Release*] because with small children it is difficult to go to the cinema very often. The programme serves to keep me in touch with the pictures until such time when I can go again' (p.4). Several respondents took a similar approach, explaining how (due to various commitments) the lull in their filmgoing was only temporary, and that *Current Release* functioned as a means through which they could maintain their interest in the cinema. As these responses suggest, the programme was seen as entertaining in itself, whether or not the viewer used *Current Release* as a guide for regular filmgoing. Yet it was clear that for certain viewers, filmgoing was becoming more of an occasional activity than had previously been the case. This was also evident in *Picturegoer*'s enquiry, including the letter which prompted the survey. The writer explained how she now only went to see the 'outstanding spectacles' that the cinema had to offer, and another reader insisted that 'Frankly, I am not now prepared to go down the road to see a mediocre film'.[79] When considered as a whole, these responses suggest a decline in indiscriminate filmgoing of the previous decades. A study in 1946 concluded that '71% of filmgoers were not influenced in their decision to go the cinema by the film being shown' (Rowntree/ Lavers, 1951, p.231). Although 1946 was a peak year where attendance was concerned (and thus not typical), when an audience for *Chance of a Lifetime* (Bernard Miles, 1950) were surveyed just under half had gone 'out of habit, drifted in to see the film by chance, or had gone because they had nothing better to do' (Harper and Porter, 1999, p.68). The surveys here are of course limited in their scope, but they do offer a contrast in this respect.

It is logical that as the number of 'indiscriminate' patrons declined, the number of occasional filmgoers increased (Harper and Porter, 1999, p.69). As Harper and Porter explain, it was the occasional cinemagoers that became the 'most important arbiters of a film's popularity. It was they who could transform the average film into a commercial success' (p.69). With regard to *Picturegoer*'s survey, some readers cited star or generic preferences as criteria which increasingly shaped their choice of film. Such factors had of course long since exerted an important influence but with regard to the 1950s audience, Harper and Porter describe how preferences became increasingly specialised (p.73). This, however, seems slightly vague and unsurprising, and attests to the limitations involved in using primarily statistical, rather than qualitative evidence. It is arguably of limited interest to comment on the existence of these trends without exploring the factors and influences behind them. Given their interest in changing filmgoing patterns and preferences at this time, it is unfortunate and - from the perspective of the cinema programme - surprising that Harper and Porter make no reference to television. As Stafford argues in his analysis of exhibition trends in 1950s British film culture, 'more audience attention might be paid to the conditions under which cinema-goers... made decisions about what to see...' (2001, p.110), a space in which television was becoming increasingly important.

If certain filmgoers were becoming more selective about the films they would pay to see, the cinema programme could play a role in this respect. As a viewer from Plymouth explained, 'It would be from the television screen that [he] chose the films [he] wanted to see' (p.4), and according to a cinema manager from Rhyl, the features which appealed to patrons went straight to the top of their 'must see it!' list (p.6). Several respondents recalled films that they had seen as a result of *Current Release*. (These included, for example, *Worm's Eye View* [Jack Raymond, 1951], *Top Secret* [Mario Zampi, 1952] and *Mandy* [Alexander Mackendrick, 1952]. *The Planter's Wife* [Ken Annakin, 1952] was mentioned several times). A Sheffield respondent explained that 'But for the seeing of something which takes our fancy on TV we should not bother going [to the cinema] as it would be difficult to assess the type we should like from the [film] titles alone' (p.7). Presumably, this had not been a 'problem' before and it indicates how people were less willing to invest in the general habit of filmgoing as a matter of routine. If arranging to go to the cinema and paying to see a film, they wanted to ensure a certain standard of entertainment in return.

While television programmes in general may have begun to harm cinema admissions, *Current Release* played a more complex role. Typical of several responses was this viewer from Bristol who explained, 'Although we decided not to go to the cinema very much when we first had our TV set... this *Current Release* feature definitely gets us interested in some of the new films and we decide to go and see them' (p.1). In fact, although some suggested that their filmgoing had declined, a number of people claimed that they now went to the cinema more *often* since the advent of *Current Release*: 'I did not go to the cinema formerly, except on rare occasions to see a film that has been talked about... [but] since I got my TV set I have been to see five films as a result of ....*Current Release*' (p.2). Another patron agreed that, 'Of course, [*Current Release*] is a good thing for the cinema trade for it is definitely making fans of a lot of people who never thought of going to the pictures' (p.2). It was the cinema managers who were particularly interested in this trend. The manager from The Plaza in Rhyl believed that television was stimulating filmgoing:

> among people who have hitherto been non-cinemagoers. The percentage of these so
> far is small, but I am convinced that it will grow, because people who I have known
> for years to be non-cinemagoers have come specially to see a film... because they
> have 'tasted' it in Current Release

(p.7).

A Birmingham manager speculated as to the profile of this audience and believed that they were middle-aged or older patrons (p.4). With the progressive 'juvenilisation' of the cinema audience as the 1950s progressed, the 16-24 age group nearly doubled in size between 1946 and '60 (Harper and Porter, 1999, p.67). This age group, however, had long since represented the backbone of the cinema audience (Rowntree/ Lavers, 1951, p.229), and while the cinema was a mass medium, there would perhaps be a higher likelihood of infrequent filmgoers belonging to the older age groups. According to the survey, it was apparently this group who were prompted to go to the cinema more often by *Current Release*, indicating that television (and thus the cinema programme) reached a potentially wider audience.

In terms of age, the audience for *Current Release* was clearly diverse. This is at least suggested by the survey, as the BBC Viewer Research reports give no indication as to age. Many of those questioned referred to their cinemagoing as a family activity. As Geoffrey Macnab explains, when the decade began the 'film as "family entertainment" was the Rank Organisation's new goal' (1993, p.204), and with Rank's refusal to book the 'X' certificate, 'the family audience was put on a pedestal' (p.218). That filmgoing was still perceived to be a family activity in this earlier part of the decade is also suggested by *Picturegoer* which launched an attempt to 'find the typical picturegoing family' in 1953.[80] Following its nationwide search the magazine concluded that 'British families may enjoy their television but... it's clear that the happy business of picturegoing is as strong as ever'.[81] As this suggests, the renewed bid to promote the cinema as family entertainment was partly shaped by competition from television. While many of those surveyed referred to filmgoing as a family activity, they were also viewing television as a family, and used *Current Release* to collectively make their choice.

In the debates surrounding *Current Release* the principles of public service were consistently opposed to its possibilities as a promotional medium. Throughout the decade, certain viewers perceived the film programme to be nothing but 'free advertising' and unworthy of television screen time, perhaps partly responding to its reception by the British press. After the first edition of *Current Release*, a BBC Viewer Research report noted that 'One small group disliked the idea behind this series, they complained that it was "sheer advertisement"'.[82] After the second edition it was again reported that 'This idea is very popular with viewers, apart from those who object on principle to "plugs" for films...'.[83] Although such responses were representative of a minority (according to BBC Viewer Research which surveyed between 500 and 600 families), they are indicative of public resistance to advertising more generally, the techniques and influence of which were to advance considerably in the 1950s. Although we've seen nuanced but significant differences between conceptions of the 'television excerpt' and the cinema trailer, in certain ways, they could function interchangeably. A cinema manager from Liverpool insisted that *Current Release* might be described as a 'modern version of the trailer service' (p.2), and this acknowledges wider promotional strategies at this time in which, due to declining audience figures, cinema trailers were becoming increasingly mobile. *Kine Weekly* observed at the end of 1955 that 16mm trailers were 'one of the most potent aids the renter can supply' and local exhibitors had arranged for screenings in places such as hairdressers, shop-windows, and factory canteens (Burton and Chibnall, 1999, p.85). Alan Burton and Steve Chibnall discuss how 16mm trailers were never fully adopted as campaign accessories and attribute this to the competition from television at this time (p.85), although this seems contradictory as they are actually similar strategies. The cinema manager above describes *Current Release* as a 'modern form of trailer service'. Perhaps it was perceived that if certain patrons were attending the cinema less frequently, they would miss the trailers for forthcoming films. Exhibitors thus needed to attract the public's attention in places other than the cinema. The cinema manager quoted above indeed notes that the trailers on *Current Release* could 'reach a very wide audience, including many people who did not normally go to the cinema' (p.2). As a result, it functioned as a 'modern' form of home cinema which served to mediate between cinemagoing and the shift toward the domestication and privatisation of spectatorship.

Part of its usefulness as a home cinema, and thus part of its public service function, varied by region, and this factor is not insignificant in considering the reception of *Current Release*. *Today's Cinema* conducted the survey across a range of areas because the trade were concerned that *Current Release* was of less use to provincial, or Northern filmgoers. This was because the further viewers lived from London, the later the release date of the films. Radio was more flexible in this respect. Although the BBC used a main national transmitter, there were also those which functioned on a regional basis. The BBC broadcast such features as *Northern Movies*, *Films in the West* and a series entitled *Provincial Release*. They originally intended to incorporate this regionalisation into *Current Release*, but this evidently proved to be more difficult than the BBC anticipated. The survey suggested that certain areas waited some time to see the films previewed on *Current Release*. (While the release dates in Bournemouth, for example, coincided with the films shown, Newcastle viewers claimed to wait 'months' before they appeared on their local circuit). Furthermore, 1952-3 was still early in the geographical spread of television, and certain transmitters were yet to open.

While such regional concerns relate to the public service functions of the review, institutional debates largely centred on the issue of the *critic* in this respect. Not unlike the BBC and the film industry, the viewers who perceived the programme to be 'sheer advertising' emphasised its lack of critical autonomy. A student from Southsea suggested that *Current Release* could be improved by the 'use of more criticism as against mere trailers' (p.2), although he still found the programme a 'useful guide'. Another respondent suggested that Fitzgerald 'should have scope for criticising the films... instead of always eulogising them' (p.3), and a BBC Viewer Research report had noted that 'some suggested the adoption of a more evaluative and critical standpoint'.[84] In this respect, we should remind ourselves of the viewer's prior experience of radio which likely shaped their expectations of *Current Release*: listeners were used to hearing reasonably critical reviews of the latest films. On the whole, however, the absence of a critical standpoint was rarely acknowledged, and this indicates that the institutional debates which surrounded the programme are not necessarily indicative of its possible reception. The BBC perceived freedom of criticism to be central to the public service function of the review; its status as a useful guide for the prospective filmgoer. For many viewers surveyed, this did not in fact hold true. As a viewer from Liverpool explained, *Current Release* 'impresses us more than do film criticisms and reports in the newspapers' (p.2), and a Birmingham respondent explained that 'It... enables[s] me to select films... as I never read reviews' (p.6). These responses suggest that viewers used the excerpts to judge *for themselves* whether they wished to see the film. While the survey offers a wide variety of responses and emphasises how *Current Release* was used in a multitude of ways, the most recurrent view was that the programme was a 'useful guide'. Of the eighty responses published in the survey, 91% explained that after viewing *Current Release* they made a note (often written), of specific films to see. If loosely defined, this attests to its complementary relations with both public service and promotion and highlights the problems involved in seeing these aspects as simply opposed or contradictory. This is succinctly expressed in the term *consumer guide* which, after all, is what the cinema programme essentially was.

The reception of *Current Release* was contradictory and varied, and the available evidence cannot be seen to be representative of the audience as a whole. However, it nevertheless provides insight into the consumption of the cinema programme at this early stage. *Current Release* was popular with regular, occasional, and so-called 'noncinemagoers', and frequency of attendance shaped the way the programme was used. This suggests that it is problematic to draw a simple correlation between declining cinema admissions and the increasing number of television viewers. The cinema programme fostered a unique dialogue between these activities with complex, shifting and often surprising results. The traditional perception of this period is one in which filmgoers are increasingly 'lost' to television, but this assumes that film and television existed autonomously and perpetuates an image of hostility and rivalry. If we are to understand the emerging shifts in leisure at this time, particularly with regard to how they were experienced, it is important to explore precisely the points where film and television interact. Whether used by regular, occasional, or 'noncinemagoers'- as a weekly guide or as pure entertainment - *Current Release* was part of the 'modern' way to experience the cinema in 1952. In terms of reception, there is less available evidence on the subsequent cinema programmes, and responses to *Current Release* cannot speak more generally for *Picture Parade* or *Film Fanfare* in 1956-8. Yet perhaps as filmgoing continued to decline, the cinema programme became an increasingly important 'link' with British film culture. The evidence of audience response represents tantalising traces of a complex interaction between viewing, filmgoing and the cinema programme which this chapter has only begun to explore.

It is clear that to dismiss *Current Release* as simply an attempt to persuade the film industry to relinquish their feature films denies the cinema programme its own institutional history, existence and reception. When *Current Release* ended its run in March 1953 it was regarded as a considerable success by both the film industry and the BBC, and available evidence points to its popularity with viewers. A venture which had in fact begun as an experimental eight-week project had developed into a popular and established form of television programme. That said, although general coverage of current films and stars continued to appear on magazine or entertainment programmes (in particular the topical *In Town Tonight* [BBC, 1954-

6] and the star profile series *Filmtime* [BBC, 1954-62]), there were no regular film review programmes for the next two years. The BBC's monopoly of broadcasting, however, was soon to be disturbed. In 1955 commercial television emerged and it was immediately clear that it was not to ignore the possibilities of the cinema programme. It was soon clear, however, that the BBC would need to fight back with something more ambitious than *Current Release*.

Notes

1. *Kine Weekly*, 24 January, 1952, p.2.

2. *Today's Cinema*, 21 January, 1952, p.2.

3. 5 February, 1952. Meeting between film industry and BBC representatives. T6/104/2.

4. 26 September, 1951. 'Current Release notes', prepared by W. Farquarson Small. T6/104/1.

5. All information for this part of the analysis is taken from scripts 1 and 2.

6. 26 September 1951. 'Current Release notes' prepared by W. Farquarson Small. T6/104/1.

7. 26 September 1951. 'Current Release notes', W. Farquarson Small. T6/104/1.

8. *Current Release* script, no.1,

9. *Radio Times*, 26 August, 1952, p.22

10. 22 November 1948. Howard Thomas to Cecil McGivern. T6/14/1.

11. 1 February 1952 Memo by P. H. Dorte. T6/104/2.

12. i *Kine Weekly*, 21 February, 1952, p.4.

13. *Kine Weekly*, 7 February, 1952, p.8.

14. *Current Release* script no.3.

15. *Kine Weekly*, 21 February, 1952, p.4.

16. *Daily Film Renter*, 4 February, 1952, p.3.

17. 19 February 1952, BBC Viewer Research report on *Current Release*.

18. *Today's Cinema* survey, 5 January 1953, p.5.

19. *Picturegoer*, 16 May, 1953, p.3.

20. *Picturegoer*, 16 May, 1953, p.3.

21. Susan was referring here to the role of Marjorie Clarke (Pauline Stroud) in the British comedy *Lady Godiva Rides Again* (1951). Although aspiring to a career in film, Marjorie finds that it is not as glamorous as it may first appear. After a series of disappointments she is forced to take employment in a risqué French theatre revue posing as the nude Lady Godiva on her white horse. Susan perhaps understood the film to suggest that there were worse ways of earning a living than secretarial work.

22. *Current Release* script, no.11.

23. *Picture Parade* script, no 1.

24. 3 February, 1952. File 3a, BBC press cuttings.

25. 27 March, 1952. McGivern to Small. '*Current Release:* Wednesday March 26, 1952'. T6/104/2.

26. To take a typical example, episode fifteen included five films which began with the 28th *Tarzan* picture, *Tarzan's Savage Fury* (Cy Endfield, 1952). This was followed by the British RAF film *Angels One Five* (1952), and then the American Navy musical, *Skirts Ahoy* (1952). Next came the British comedy *Something Money Can't Buy* (1952) and lastly the historical costume drama, *Carrie* (William Wyler, 1952), an adaptation of Theodore Dreiser's novel.

27. *Current Release* script, no.7.

28. *Current Release* script, no.9.

29. It is obviously not possible to ascertain this from the scripts, and my analysis here is based on Fitzgerald's later role in the existing editions of *Film Fanfare*.

30. 19 March 1951, McGivern to Dennett. T6/104/1.

31. *Kine Weekly*, 4 January, 1952, p.2.

32. *Kine Weekly*, 21 January, 1952, p.2.

33. *Today's Cinema*, 21 January, 1952, p.2.

34. *Kine Weekly*, 24 January, 1952, p.2.

35. *Kine Weekly*, 2 February, 1952, p.4

36. *British Kinematography*, 1953. The article was a transcript from a paper given at a meeting of The Theatre Division on 6 March, 1953.

37. 15 April 1952. T6/104/2.

38. 31 January 1952 Small to McGivern. T6/104/2.

39. *Current Release* script no.8.

40. *Picturegoer*, 1 March, 1952, p.11.

41. *Kine Weekly*, 24 January, 1952, p.2.

42. 26 September, 1951 Del Strother to McGivern. T6/104/1.

43. *Current Release* script no.1.

44. Script for *The Week's Films*, 3 September, 1944.

45. *Current Release* script no.7.

46. *Variety*, 22 September, 1952.

47. *Current Release* script, no.11.

48. *Daily Express*, 17 March, 1973. BFI press clippings folder on cinema programmes.

49. *Current Release* script no.5.

50. 18 March, 1952. Meeting between film industry representatives and the BBC. T6/401/2.

51. 1 July, 1952. Meeting between film industry representatives and the BBC. T6/104./2.

52. 3 October, 1952. Small to McGivern: 'Current Release: Star Interviews'. T6/401/2.

53. As above.

54. *Radio Times*, 28 December, 1952, p.66.

55. 12 December, 1952. 'Memo on *Current Release*'. T6/104/2.

56. 21 December, 1952. BBC press cuttings file 3A.

57. See *The Television Annual of 1952*, p.45-53.

58. *Picturegoer*, 9 February, 1952, p.5.

59. As above.

60. *Today's Cinema*, 5 January, 1953, p.1.

61. The survey is available in press cuttings box P657, WAC.

62. Unless otherwise stated, all the following responses are from the trade survey, and are only followed by the relevant page number.

63. *Picturegoer*, 8 March, 1952, p.18.

64. BBC Viewer Research report on *Current Release,* 4 February, 1952.

65. 1 April, 1953. Notes of a meeting between BBC and film industry. T6/104/3.

66. See Audience Research Bulletins, Television, February 1950-June 1953, R9/4. *Picture Page* received a reaction index of 60, while *Current Release* was usually in the 70s. Many programmes received ratings in the 40s and 50s.

67. *Picturegoer,* I March, 1952, p.7.

68. *Daily Mirror,* 5 June, 1952.

69. See discussion in the correspondence 29 November 1957, Madden to British Lion, T6/360.

70. This information comes from an extensive survey of housewives conducted by the Newspaper Society between September 1957 and June 1958.

71. 31 January, 1952. Small to McGivern. T6/401/2.

72. *Picturegoer,* 1 March, 1952, p.7.

73. As above.

74. *Picturegoer,* 1 March, 1952, p.7.

75. *Picturegoer,* 19 March, 1955, p.5.

76. *Picturegoer,* 11 August, 1953, p.3.

77. *Picturegoer,* I March, 1952, p.7.

78. *Picturegoer,* 1 March, 1952, p.7.

79. *Picturegoer,* 1 March, 1952, p.7.

80. *Picturegoer,* 7 November, 1953, p.25.

81. *Picturegoer,* 28 November, 1953, p.9.

82. BBC Viewer Research report on *Current Release,* 4 February, 1952.

83. BBC Viewer Research report on *Current Release,* 19 February, 1952.

84. BBC Viewer Research report on *Current Release,* 4 February, 1952.

Chapter Four

The Cinema Programme in the Age of Competition:
Picture Parade and *Film Fanfare 1956 - 8*

The advent of commercial television in 1955 marked the beginning of a period of intense activity where the cinema programme was concerned. While the BBC continued to play an important role in the development and circulation of the genre, it is clear that the emergence of a second channel increased the opportunity for television's coverage of film culture. Relations with the film industry - while still fractious and contradictory - became more routinised, and on a wider level, the cinema programme enjoyed a cultural and institutional significance it has not enjoyed since. While these programmes were clearly rooted in the heritage of their predecessor, *Current Release*, the context of the new competitive television environment influenced the focus, aesthetic and format of the genre in which, increasingly driven by coverage of the stars, they indeed become 'dazzled by glamour' (the phrase later used by Buscombe to criticise the cinema programme's 'trivial' construction of film culture (1971, p.62)). However, it is in many ways precisely this dazzling 'excess' and exuberant celebration of the cinema that is central in understanding the historical specificity and cultural significance of the cinema programme at this stage.

Enter ITV

In 1949 the Beveridge committee recommended the continuation of the BBC's monopoly of broadcasting but as the 1950s progressed, the Conservative government reopened the monopoly debate and pressure grew for the introduction of an alternative service. Advocates of commercial television offered several reasons why the current broadcasting infrastructure should come to an end. There were many critics who perceived that, under the BBC, the development of television had been held back by its economic and institutional subservience to radio. The threat to the monopoly also emerged from changes in the socio-political climate, what Crisell describes as the campaign to 'let the people decide for themselves', a 'renewal of the process of democratisation' that had been gathering pace before the war (1997, p.77). Other challenges were economic. The growth in prosperity in the 1950s meant a rise in production and a corresponding demand for advertising outlets - in which respect television could provide a valuable function (Crisell, 1997, p.78). After the Television Act of 1954 it was decided that commercial television was to be supervised by the Independent Television Authority (ITA). The ITA was to allocate a number of regional franchises to programme companies (the 'contractors'), and they would comprise a single network. The regional concept was a response to Beveridge's complaint about the excessive 'Londonisation' of broadcasting and the ITA had envisaged a service which would cater to the 'special tastes and outlook' of its viewers (Sendall, 1982, p.308). However, it soon became clear that networking, the exchange of programmes between contractors, was economically necessary for ITV's survival, minimising the possibilities of a regional inflection to commercial television. Networking also reduced the competitive relations between the contractors so that the only real competition was between the programme companies and the BBC (Crisell, 1997, p.86).

Commercial television rapidly depleted the BBC's audience and a recurring argument in many histories is that the BBC were unable to compete with the innovative and populist ITV. This struggle is seen as symptomatic of a more general cultural shift in which the BBC's 'paternalism' was deemed to be increasingly inappropriate (as Crisell's conception of 'democratisation' above suggests). The audience ratings were a topic constantly discussed in the press at the time and certainly indicated the popularity of the new channel. Although methods of calculation were subject to question (see Briggs, 1995, p.21), in 1957 the BBC's *own* research indicated that in homes with a choice, ITV attracted 72% of the viewing audience (Briggs, 1995, p.20). In the same year, splits as dramatic as 79:21 were reported in ITV's favour (Crisell, 1997, p.88). As a result, it is not surprising that the BBC regarded the new competitor with much hostility. Gerald Beadle was made Director of BBC television in 1956, and his book *Television: A Critical Review* (1963) explained the BBC's institutional logic at this time. Both he and the Director General of television understood that ITV would initially draw the larger audience, and the BBC was prepared to drop to a 40% share. Beadle assumed, however, that this margin would gradually be reduced and that ITV would come out and compete with the BBC, gradually 'raising their standards to [the BBC's] level' (Beadle, 1963, p.79). This reiterated the view expressed by the supporters of the BBC's monopoly (and is still expressed today in the context of the increasing de-regulation of the television landscape), that competition would drive out quality. Peter Black suggests that 'The BBC continued to prepare its schedule as though the competitor did not exist' (1972, p.134), and Bernard Sendall describes how the ITV companies awaited the BBC's 'counter-blow, and were puzzled when it did not arrive' (1982, p.112). Such arguments are often supported by the suggestion that 'Independent Television provided a lively counter to the stuffiness of the BBC' (Stokes, 1999, p.34), or that ITV won the audience not only with shifts in programme content and scheduling, but 'with a friendly style of presentation that the BBC had always found elusive' (Black, 1972, p.116, see also Sendall, 1982, p.325).

It is true that the *TV Times* and the *Radio Times*, attached to ITV and the BBC respectively, provided a striking contrast, and likely a very different reading experience for the viewer. The *TV Times* focused on personalities and reader competitions, and clearly differed from the BBC's more functional and informative publication. While these intertexts shaped the reception of television, it is perhaps more hasty to make generalised claims about programme material. ITV certainly increased the amount of popular and light entertainment features on British television, and often cited are its American telefilm series, quizzes and variety shows. Yet it is difficult to see how confident assertions about the details of programmes (such as modes of address) have been formed when comparatively little of British programming at this time has been analysed in detail (partly due to the paucity of available sources). The issue of channel comparison appears to be particularly crucial in these early years of co-existence between the BBC and ITV, and the wider development of the cinema programme is firmly situated within this period. As noted in Chapter Two, from this perspective it is worth acknowledging that, rather than automatically accepting or replaying existing arguments, a key implication of the 'local' analysis of specific genres should be to reflect back on the arguments and discourses which structure our 'macro-overview of broadcasting history' (Jacobs, 2000, p.9). As generic studies of early television programming increasingly emerge, these issues are central to debates about the critical and methodological construction of television history, and the questions we ask when approaching this field. For example, what do we do if a genre doesn't seem to 'fit' with these wider institutional histories? Is it an anomaly, or does it indicate the need to question the certainty with which such histories have been drawn? There are certainly differences between the BBC's *Picture Parade* and ABC's *Film Fanfare*, but situating these is no simple matter. The key issue of *how* these differences might be conceived and perhaps most crucially, the ways in which they might be contextualised, cannot necessarily be reduced to an overriding comparison between the channel's conception of the audience, or 'general' styles of presentational address. Equally, it is worth emphasising that the series also displayed as many similarities as differences.

In terms of questioning some of the wider perceptions of the BBC's position and strategy at this time, Andrew Crisell comments that:

> The BBC's programming was by no means weak. Indeed most of the memorable programmes which date from this era are the Corporation's. Part of the problem lay in the way they were scheduled. Under the Director General Sir Ian Jacob the policy was partly to compete with ITV, to match like with like, but true to its old public service philosophy of providing a range and catering for minorities, the BBC also tried to offer contrast by scheduling some of its more serious output at peak viewing hours
>
> *(1997, p.88).*

The cinema programme confirms the attempt, as Crisell suggests, to 'match like with like'. This was clearly a time of uncertainty and change for the BBC and it was not until the 1960s and 1970s that the channels settled into a 'cosy duopoly' which was to characterise British broadcasting for some time (Scannell, 1990, p.21). In some respects, however, without this structure in place, competition between the channels could be *more*, rather than less, pronounced. It is equally important not to over-emphasise the populist, commercial identity of ITV. This is particularly so when the early years of the channel (in comparison with the BBC) remain relatively unresearched, and as Paddy Scannell emphasises, the terms under which commercial television was established in Britain made it part of the public service system from the beginning (1990, p.17).

ITV was initially restricted to the London area, and transmissions began on 20th September 1955. Programme hours were extended and both channels began to broadcast a 49-hour week - with the BBC also investing an extra £30,000 a week in television programming (Black, 1972, p.130). It was not simply the range and hours of programming that expanded as the television audience had greatly increased, and even the most cursory glance at the discussion of television in the *TV Times* or *Picturegoer* attests to the ways in which it was increasingly penetrating everyday life and cultural consciousness. By 1956 the ownership or rental of a television set was passing through and 'out of the stage of being a marker of status within the working class. TV was on the way to becoming a standard feature of every home...' (Corner, 1991, p.6). Unlike *Current Release*, *Picture Parade* and *Film Fanfare* were to reach an increasingly mass audience. (In 1956 television licenses totalled 5,739,593, as compared to 1,449,260 in 1952) (Briggs, 1995, p12).When ITV first began, its audience was limited to the London area. Even within this region, only 33% of viewers had sets which could receive the new channel (Crisell, 1997, p.87) and, as Briggs reminds us, on the night that commercial television began, there were still more radio listeners than viewers (1979, p.9).

Nevertheless, while its geographical penetration may have been limited, the BBC were well aware of what a memo titled 'The [ATV] plans for the London Weekend'.[1] Featuring a list of the programmes scheduled in ITV's first weekend of transmission, this included a cinema programme called *Movie Magazine* (ATV, 1955-6). In the London area, Associated Rediffusion (AR) provided the weekday programming, and Associated Television (ATV) the weekend. *Movie Magazine* was produced by ATV and screened on Sunday evening (7:45-8:00pm), and given that *Current Release* was screened on a fortnightly basis, *Movie Magazine* could claim to be the first *weekly* film-review programme on British television. ITV was to prompt a shift in television scheduling as it encouraged viewers to form habits on a weekly basis (Sendall, 1982, p.325), and from this point on, the cinema programme followed this pattern. Like *Current Release, Movie Magazine* was presented by John Fitzgerald and for viewers who had a set in 1952, he must have represented a familiar face. (After *Current Release* ended in 1953, Fitzgerald wrote a weekly review column in *Picturegoer* (also entitled *Current Release*)). While

Fitzgerald had only been contracted for *Current Release* rather than employed by the BBC on a regular basis, he was one of many presenters, actors and staff who moved over to ITV. As the *TV Times* later emphasised, however, it was through *Current Release* that Fitzgerald 'made his name' in television[2] and - perhaps making for a degree of continuity between the channels where the coverage of film was concerned - he was to feature regularly in the cinema programmes on ITV.

Offering the viewer 'news and scenes from popular films',[3] *Movie Magazine* was only fifteen minutes long, although such short programmes were not uncommon at this time. *Movie Magazine* was actually quite popular as it appeared in the top twenty ITV programmes in October 1955, watched by 66% of the London viewers.[4] It gradually increased in length and by December 1955 it was half an hour long, screened at 3:30pm on a Sunday afternoon. But the wider rise in film programming at this time also had much to do with the influx of profile features, programmes focusing on a particular star or director. In media treatment of the arts, the concept of 'the artist profile' has a long history (Kerr, 1996), and was certainly already established in print media and radio. Some of the earliest film-related programmes on BBC television in the late 1940s were organised around directors (such as Griffith and Eisenstein), 'suggestive ... of cinema being taken seriously as an art form by television at ... an early stage...' (Kerr, 1996, p.136). However, as Kerr notes, the most prominent type of film profile features on television - at least in terms of its popular coverage - have historically been those focusing on the star (Ibid), and this was evident from an early stage. In January 1956 ATV began its fifteen-minute programme *Portrait of a Star*, an early edition of which focused on Dirk Bogarde.[5] Although initially scheduled at 7:45pm, *Portrait of a Star* ultimately found its regular slot at 10:30pm on a Sunday night, with John Fitzgerald providing the commentary to accompany the clips. AR was soon to offer its own profile series called *Close-up* screened at 9:00pm on a Wednesday night, and it continued to run for several years. AR also produced the fifteen-minute *The Other Screen* as part of the daytime women's programme, *Morning Magazine* (1955), and at 11:45am on a fortnightly basis, female viewers were introduced to a discussion of 'films and the people who make them'.[6] At other times, it was simply described as a 'review' of film, and the incorporation of film into women's programmes was a strategy previously adopted by the BBC in both radio and television. The mid-1950s also saw the advent of cinema programmes for children (at least on the BBC). Scheduled at 5:30-5:45pm and directly following the period designated as 'children's television', *Junior Picture Parade* (BBC) began in 1956,[7] and was later followed by *Filmland* (BBC) (5:15-5:35pm).[8]

Once further ITV transmitters were operational, coverage of film culture increased. In February 1956, ITV began transmission in the Midlands, where ATV would provide programming on weekdays, and a newcomer, ABC Television, on weekends. The Northern audience had to wait until May 1956 to receive ITV, where

Granada was responsible for weekdays, and ABC the weekends. The opening of ITV in the Midlands was heralded as the 'big' second phase of the new channel, announced with much excitement in the *TV Times*. This equally marked the emergence of ABC's large-scale series, *Film Fanfare*, and only six weeks later the BBC's rival *Picture Parade* appeared. The battle was on.

Establishing Relations: The BBC, ABC and the Film Industry

It is unclear whether *Picture Parade* was specifically produced as a direct response to ITV and its increasing coverage of film culture. McGivern claimed that the BBC had been planning a cinema programme for some time,[9] but there had been no regular designated film series since the end of *Current Release*. It also seems no coincidence that *Picture Parade* emerged when it did. The television audience was clearly now a site of competition; it had to be courted and enticed by the different programmes on offer (Crisell, 1997, p.88). Although its audience was declining, the cinema was still a very popular medium and, as a result, it is not difficult to see why film became an area of competition between the channels. Despite the claims of the press that television was at the mercy of the film industry where these programmes were concerned, the BBC and ITV were benefiting from the coverage of the cinema as much as the film industry themselves.

Throwing doubt on the claim that the BBC prepared their schedule as though 'ITV did not exist', the BBC were aware of ABC's plans for *Film Fanfare* well in advance of its first transmission.[10] Weeks before *Film Fanfare* began, the BBC's Cecil Madden informed McGivern that:

> I learned that Howard Thomas [Managing Director of ABC TV] has notified ATV that he plans a 1 hour film magazine on Sundays, 3-4pm, and that in return [for other regions showing Film Fanfare] he will network Sunday Night at the London Palladium and perhaps others. It means that [Movie Magazine] ... on ATV will have to move. This is the hour I wanted for... Picture Parade.[11]

Film Fanfare was initially a Sunday afternoon programme and later became a Saturday night feature screened at 10:00pm, while *Picture Parade* was ultimately screened at 9:15pm on Mondays. Yet the specific detail of their scheduling is not as important here as the sense that they were clearly understood in their general planning, development and conception, as *competing* series from the start. In their planning, scheduling and launching of a cinema programme at this time, it seems clear that the Corporation did not intend to be 'beaten' in an area which they had pioneered. McGivern, now Deputy Director of Television, was clearly aware of the importance of light entertainment in the era of competition, as well as the role of the cinema programme within this scheme. Staff were advised that *Picture Parade* was to be regarded as a priority where film extracts were concerned and McGivern insisted that any other features (the profile programmes, for example) must now be 'special and good' if they were to be screened.[12] Standards, it was clear, were to be raised.

It is in relation to *Film Fanfare* that the move of film companies into television at an institutional level becomes important. As discussed in Chapter One, British film companies such as the Associated British Picture Corporation (ABPC) and Granada (which owned only cinema chains) submitted successful bids for ITV franchises, becoming programme contractors in 1956, and in some respects, this significantly altered the relations between the film industry and television. As indicated, the film companies' move into television was regarded as controversial by the British film industry. ABPC clearly understood this - hence in part explaining Thomas' repeated claim that the company had gone into television to 'boost cinema admissions'[13] - and there is a sense in which the institutional base of ABC TV shaped the textual construction of *Film Fanfare* in ways which differentiated it from the BBC's *Picture Parade*. What became ABC TV was part of the Associated British Picture Corporation (ABPC), one of the most powerful organisations in the British film industry from the late 1920s to the late 1970s (Porter, 2000, p.152). This power emerged from the fact that, like the Rank Organisation, ABPC was a vertically integrated company which not only produced films at Elstree studios, but also owned the distribution outfit Associated British Pathé and the well-known ABC cinema chain. Vincent Porter (2000) has considered the films produced by the company in the 1950s and aspects of their production regime; Allen Eyles (1993) has researched the ABC cinema chain, and Howard Thomas' (1977) biography (first managing director of ABC TV and formerly producer-in-chief at Associated British Pathé) discusses the company's move into TV, as does Bernard Sendall's (1982) volume on the emergence of commercial television. What is lacking is an investigation into how the various aspects of ABPC interacted and the extent of their convergence. This is particularly so in terms of the relations between ABC TV and the ABPC's film interests, a synergy which is key in understanding *Film Fanfare*. In this respect, *Film Fanfare* acts as a crucial indicator that what was perhaps most interesting about ABPC in the 1950s was its identity as a company at the forefront of a changing media environment.

In terms of ABPC film production at Elstree studios, the company produced or co-financed some eight pictures a year, fulfilling the quota of British films for the ABC cinema chain (Porter, 2000, p.151). Although there were exceptions, budgets were kept low and few of ABPC's pictures achieved critical or commercial success (Ibid). ABPC had its own stable of stars, and while the British film industry clearly never operated a star system comparable to that of Hollywood, ABPC and the Rank Organisation were the two British companies to sign stars on long-term contracts (Macnab, 2000, p.173). (British stars contracted to ABPC at this time included names such as Richard Todd, Sylvia Syms and Janette Scott). Vincent Porter laments that by 1958, the company filled Elstree with 'television productions and restricted its film investments to comedies featuring television stars' (2000, p.159), yet this is precisely where the interaction between the company's wider media interests seem important. ABC TV had been established for two years by this time and evidence suggests that Associated British stars were regularly used in television and vice versa.[14] Elstree studios had been used for television production for some time, with ABC TV offering an early example of a successful British telefilm series, *Robin Hood* (1955-60). ABPC produced approximately one-third of the films it distributed in the 1950s and the scope of its distribution arm was significant, operating in both Britain and overseas (Thomas, 1977, p.120). Associated British Pathé also produced the twice-weekly newsreel *Pathe Gazette*, as well as the weekly magazine *Pathé Pictorial*. In terms of exhibition, ABC was a major force and in the mid-1940s it owned the largest single circuit in Britain. In 1955 ABC TV was added to the corporate empire. ABPC displayed a keen interest in joining Independent television well before the Television Act of 1954. As Bernard Sendall notes, ABPC:

> had everything going for them - modern premises at Elstree, all the technical resources of film production, a subsidiary called Associated British Pathé ... in terms of studios .. and entertainment experience they could be considered the strongest applicant

(1982, p.82-3).

In 1955 ABPC's annual report indicated that refurbishment of the company's cinema chain had demanded huge economic resources, and they were forced to drop out of the running. However, when it was discovered that one of the franchises was unable to participate, ABPC again considered entering ITV. The company persuaded Warner Bros, who had held a one-quarter share in the company since 1941, to agree to the venture, and their application was accepted the day before ITV went on air (Thomas, 1977, p.125). Howard Thomas had forged relations with the BBC from an early stage, regularly contacting Cecil McGivern in the late 1940s with television programme ideas on films or filmmaking. With full co-operation of ABPC, in 1953 the BBC produced *Sneak Preview,* a one-off programme centred on the exhibition of the Associated British comedy *Will any Gentlemen?* (Michael Anderson, 1953).

'To work whole-heartedly together in the interests ... of show business': Formalising Relations

Before considering the programmes in detail, it is necessary to understand the institutional relations which underpinned them, as well as the changed position of the film industry at this time. During the production of *Current Release* the film industry had simultaneously been chasing the possibilities of Cinema-TV - a time when television's technological and cultural identity was more unstable, and when the decline in cinema admissions was less dramatic. From 1955, but particularly during the years 1957-59, the drop becomes sharper, with more than 300 million admissions lost (Stafford, 2001, p.96). John Spraos' rather apocalyptically entitled *The Decline of the Cinema* (1963) describes the second phase of decline as beginning in 1955, with 'the dramatic expansion of television among the working class. In this phase, each new set dealt a heavy blow to the cinema' (1963, p.22). Spraos, however, rejects the argument that this can be directly correlated with the expansion of ITV, suggesting that it was equally precipitated by the closure of many local cinemas, a process which gathered ground after 1956 (p.29). Either way, Stafford insists that 'Wherever the blame for decline is pinned, the key year was 1955 ... [After this time] it was clear that the cinema as a mass entertainment medium is over' (2001, p.97). While it is certainly true that this period represented a changed economic and cultural context for the cinema than earlier in the decade, this nevertheless seems a little hasty. The situation was still perceived as more uncertain in the trade press, and the notion that the cinema had had 'its day' is difficult to reconcile with the film industry's relations with television, or with the cultural texture of the cinema programme. In fact, with the competition from a new channel, the BBC was keen to further cement their relations with the film industry. In 1956, the BBC's Cecil Madden addressed the trade in the *Daily Film Renter*, suggesting that 1956 should herald a new dawn of co-operation where the cinema programme was concerned. Not to be outdone by ITV, he insisted that:

> The BBC wants to present films that are an exciting and sparkling aspect of show business ... Let us make 1956 a year in which the film industry and television will for the first time work whole-heartedly together in the interests of each other and of show business as a whole.[15]

Nevertheless, despite greater degrees of formalisation, it is possible to detect a number of heightened anxieties circulating around these relations, at least from the film industry's point of view. In the *Current Release* survey, *Today's Cinema* enquired as to whether patrons would stay in to view the series, and as television ownership increased, so did the industry's concerns about the contradictory implications of the cinema programme. Again we see how the crucial sphere of tension here is *exhibition* - both in terms of complaints from cinema exhibitors, and the role of television in providing a further outlet for film. The committee involved in liaising between the film industry and the television channels increasingly stipulated *when* the programmes could be screened, and this was part of a wider framework regulating the relations surrounding the cinema programme at this time. The Film Industry Publicity Circle (FIPC) (who were involved with *Current Release*) collaborated with the Kinematograph Renter's Society (KRS) to form a joint committee to deal with television's requests for film clips. When *Picture Parade* began, the trade press published its 'Code for Film Excerpts on TV' and initially listed the programmes to be supported by the committee - including

Picture Parade, *Film Fanfare* and *Movie Magazine*. Outside of the committee, companies could provide clips for the star-orientated programmes at their own discretion. To ensure that clashes did not occur, members were requested to inform the society of any commitments they made, and the forms for this purpose were to be supplied by the company Publicity Directors. While emphasising how relations between the media were becoming formalised in a way that had not been the case with *Current Release*, the film industry's desire for such carefully *regulated* interaction equally spoke of their defensive attitude toward television and, concurrently, their increasing concerns about the future of the cinema.

The film industry's intervention in the scheduling of the programmes basically prevented them from being screened between 5pm and 10pm, and this emerged from a broader context of debate which exceeded the parameters of the cinema programme. As discussed in Chapter One, scheduling had been a prime concern where the televising of feature-length films were concerned, and the BBC always reassured the film industry that if films were made available, they would broadcast them at a time when they might have least effect on the box office.[16] This partly explains why the film industry were so concerned about the scheduling of the cinema programmes, as *Picture Parade* and *Film Fanfare* were still an important way of experiencing 'film' on television - and hence were also perceived to be more in competition with the cinema itself than other television fare. This was emphasised by the *Daily Express* in its article 'Let's Have TV on our TV Screens', with the implication that it was 'film', rather than 'TV', that was dominating the schedules. It reported that: Competition between the BBC and ITV in bringing... the latest news and views from the motion picture world, is livening up into a tidy little tussle' and disapprovingly observed that "converted' viewers now have eight different programmes on films to choose from. Surely this is too much.[17]

From a retrospective point of view at least, what is striking is that the film industry seem to have no appreciation (or concept) of 'prime-time' in television here. In prioritising the space of the cinema within that particular temporal stretch of the day, they were also denying the possibility to maximise the audience for the cinema programme (and hence the potential trade it may encourage).

However, although with sometimes minor flouting of the rules, both the BBC and ITV appeared to respect the film industry's wishes in their desire to maintain amicable relations. What is apparent in these debates was that (as with FIDO), the film industry's stipulations were very much short-term plans for interaction with television, designed to minimise its effect on cinema admissions at that time. The *Daily Film Renter* recognised these contradictions when it boldly offered an alternative point of view. Speaking of 'the dilemma before which exhibitors stand hypnotised' - that the cinema programme both promoted yet competed with the cinema as a site for film consumption - it insisted that:

Thinking of TV in these terms belongs to 1953 not 1956... It is no longer a question of revising a program [sic] like the old *Current Release* which showed snippets of films... to whet [viewers] ... appetites for the program [sic] at their local cinema. That has had its day. We should accept the realistic fact that no one goes to the cinema seven nights a week. More important, no one watches television seven nights a week ... Our problem today is how we can co-operate with TV in such a way that the box-office ... derives benefit from the new situation.[18]

This seems to acknowledge a shift in cinema's cultural role and, concurrently, its relationship with the cinema programme. Firstly, it acknowledges changing patterns in cinemagoing. As discussed, 1956 was the first years of significant cinema closures, with Rank shutting down fifty-nine and the ABC chain thirteen (Eyles, 1993, p.78). While many local cinemas of course continued to exist, it was often these that were hit hardest by the decline and closed first, at least those that were not part of a larger chain (Spraos, 1963, p.85). In the *Daily Film Renter*'s conception, the 'local' cinema is connected with regular (perhaps indiscriminate) filmgoing, and it is this which it perceives to be in decline. However, this does not necessarily imply, as Stafford suggests, that this perception is based on the notion that the cinema's days as a mass medium are over. This particular article perhaps hints at this, but it is more concerned with the more immediate detail of changing exhibition patterns.

This may partly explain why the film industry attempted to centralise the relationship between the films previewed on the programmes and cinema exhibition (and it is worth noting here that although the closure of cinemas clearly hit the major chains, overall it increased their market share) (Eyles, 1993). The quote from the *Daily Film Renter* indicates how *Current Release* dealt with films at the level of the 'local' - although, of course, the concept of the local release could vary considerably by region. ITV made possible a regional inflection to the cinema programme (as titles such as *Midland Movie Magazine* suggest), yet the concept of networking reduced these possibilities. ABC, for example, networked *Film Fanfare*, but it used the London release as its guide. In this respect, the later programmes differed from *Current Release* which had rarely featured films only showing in London, and very seldom had it previewed a pre-release picture. *Picture Parade*, however, was to preview films on West End pre-release, while *Film Fanfare* was to feature films playing on the London Odeon and Gaumont chains, as well as its own (London) ABC circuit. This meant a good deal of repetition across the programmes, and film companies had to liaise with the television channels in order to work out which films would be previewed and when. As one publicity manager noted, 'The whole thing has got so complicated that it is almost "bad business" for [a film] ... not to be seen on television'.[19]

With a Fanfare and a Parade: The Consolidation of a Genre?

A key aspect of both programmes were the presenters, and *Picture Parade* was presented by the popular duo of Peter Haigh and Derek Bond. In many ways they were similar to John Fitzgerald: male, in their 30s, with no specialist knowledge of film, and they attempted to adopt a relaxed and amiable mode of address. The dapper Peter Haigh with his immaculately trimmed moustache and carefully modulated voice was a well-known television personality on BBC television (Figure 1). Although it is difficult to find evidence of this at the time, when Haigh died in 2001, newspaper columnists foregrounded his status in the 1950s as 'television's most eligible bachelor', a tag he forfeited in 1957 when he married a starlet from the Rank Organisation, British actress Jill Adams.[20] As well as *Picture Parade*, Haigh's regular jobs included hosting *Top Town* (a series of weekly talent contests for amateur and semi-professional variety entertainers) (Vahimagi, 1996, p.149) and *The Week Ahead* (a ten-minute trailer programme). Just after *Picture Parade* began he presented a new radio feature called *Movie-Go-Round* and was involved with other radio programmes such as *Housewives' Choice*[21] - perhaps a further indication of his perceived appeal to a female audience. So extensive was his involvement with BBC broadcasting that *Picturegoer* dubbed him 'Perpetual Peter', with the implication that his appearances were a little *too* frequent.[22] The list of Haigh's appearances does not sound excessive today, and the magazine's response

Figure 1: Peter Haigh, presenter of Picture Parade (1956, edition 18a)

is a reminder of the comparatively limited choice of channels and programming, and the impact of this context on the cultural development of the TV personality (see Medhurst, 1991). Haigh co-hosted the programme with Derek Bond (Figure 2), who began his career as a film actor, and continued to work in films while presenting *Picture Parade*. He made his screen debut in Ealing's prisoner of war drama *The Captive Heart* (Basil Dearden,1946), then went on to star in the studio's historical costume picture, *Nicholas Nickleby* (Alberto Cavalcanti, 1947), and *The Loves of Joanna Godden* (Charles Frend, 1947). In terms of his film career, an appeal to a female audience is again suggested by *Picturegoer*'s subsequent description of Bond as British cinema's 'first postwar heart-throb'[23] (and it is an interesting question as to whether his role in *Picture Parade* was seen as a chance to rekindle or draw upon this status). In 1949 his film career took a turn for the worse when he transferred his contract from Ealing to the Rank Organisation, as his films were not a great success.[24] In the six years that followed he decided to work in theatre and on BBC television plays until he was asked to act as interviewer for the BBC's afternoon programme, *Filmtime*, and the offer of *Picture Parade* soon followed. In 1957 *Picturegoer* ran the headline 'TV ends those six lean years' and reported that Bond was about to appear in *Rogue's Yarn* (Vernon Sewell, 1958), 'his third starring picture in a matter of months'.[25] The magazine claimed that there was now 'an array of movie offers awaiting him' and that the reason for this was new

Figure 2: Derek Bond, presenter of Picture Parade (1956, edition 18a)

visibility on television and in particular, *Picture Parade* (also indicative of the increasing crossover of stardom between television and film). At the age of thirty-six, the weekly series had increased his popularity 'and captured a whole new public'. Bond explained how 'It was a complete reversal for me - interviewing film personalities instead of appearing as a star. But I had learned not to hang on to the shreds of a career'.[26] Interviewing on *Picture Parade* is clearly constructed here as a step down from acting in films, a hierarchy shaped by conceptions of film versus televisual fame (discussed in Chapter Five).

Complementing its more varied format, *Film Fanfare* employed several presenters, some of whom also had a background in film. Joining the programme four months after it began, John Fitzgerald undertook the task of previewing the films, just as he had in *Current Release*. John Parsons was the programme's 'roving reporter' and in the capacity of interviewer, he had occasionally worked on *Current Release*. Then there was Macdonald Hobley, an important figure on *Film Fanfare* and a key early television personality. Initially working on radio, Hobley was the BBC's first post-war television announcer and also presented the magazine programme *Kaleidoscope* (BBC 1946-53) before joining ITV in 1956 to become a key face in entertainment programming[27] (Figure 3). Hobley's significance as an early representative of television presenting is suggested by the fact that he played himself in Ealing's satire about the evils of television, *Meet Mr. Lucifer* (Anthony Pelissier, 1953), and in 1954 he was voted 'TV Personality of the Year'.[28] He also played very minor roles in films (such as *Man of the Moment* [John Paddy Carstairs, 1955]). Hobley was joined by Peter Noble, who undertook the role of presenting *Film Fanfare*'s 'gossip spot' and interviewing the stars. In addition to both presenting and writing film programmes for BBC radio, Noble was a well-known show business journalist at this time, and was also to become the author of several books on the film industry, including the first study of racial discrimination in Hollywood - *The Negro in Films* (1948).[29] In the 1950s, Noble was an associate producer on a number of British films, and although in apparently minor parts, he had acted in British features such as *The Bells Go Down* (Basil Dearden, 1943), and *It's that Man Again* (Walter Forde, 1942). This background would seem to indicate the most specialist knowledge of film, although it was really his involvement with 'showbiz' journalism that was central in his contribution to *Film Fanfare*. Like Derek Bond, the other presenters - Paul Carpenter and Peter Arne - had had more extensive careers in film. Their films are too numerous to detail here (with each appearing in more than 30 pictures), but they had acted in both British and American films, although not usually in starring roles. Interestingly, however, Arne was contracted by the ABPC from 1956,[30] thus presenting the programme while also appearing in the company's films. Also of significance here is that both Carpenter and Arne were Canadian and their accents added to the relaxed mode of address pursued by television and, according to conventional accounts, particularly by ITV. Given its generalised reputation as offering a more popular and appealing

Figure 3: MacDonald Hobley and his 'quiz' wheel (1956, edition 7)

address to television's increasingly mass audience, the use of Canadian presenters (apparently intending to 'stand in' for American and thus signify 'Hollywood') was perhaps an attempt to soften the strong middle-class address characteristic of television at this time, although the other *Film Fanfare* presenters (Parsons, Noble, Hobley and Fitzgerald) epitomise this tone in a similar manner to the BBC's Haigh and Bond. Nevertheless, with their desire to promote 'proper' standards for the enunciation of the English language, it is perhaps difficult to imagine the Canadian presenters on the BBC at this time.

Although their involvement with the film industry was not explicitly foregrounded by the programmes, the use of film actors was perhaps designed to give the impression of 'inside' knowledge - that the presenters (and thus the discourse of the programmes) emanated from *within* the film industry. To a certain extent, this idea was pursued by both programmes in a way that differentiated them from *Current Release*, a strategy which, as will be discussed, further problematised their relationship with film 'criticism'. Although the set of *Current Release* had attempted to construct this close relationship, to convey the impression that Fitzgerald was presenting the programmes from 'inside' a cutting room, for example, this had been less evident at a discursive level. It essentially provided a commentary *on* the films, which created a certain distance. The opening titles of *Picture Parade* acknowledged that it was presented with 'the full co-operation of the film industry' (a phrase also used in the listings for ITV cinema programmes in the *TV Times*). This was part of a conscious effort to promote 'good' relations with the film industry, while it also functioned to legitimise the programmes' perspective on film culture. Although *Picture Parade*'s announcement in its opening titles, particularly the use of the word 'co-operation', necessarily recognised film and television as *separate* media which came together for the programme, *Picture Parade* and *Film Fanfare* actually blurred any distinction between the film industry and the televisual space of the programmes. In *Picture Parade* the presenters would make such comments as '[We] spend a lot of time in Wardour Street, the heart of the British film industry, and ... hear a lot of interesting gossip there...'[31] Here, Derek Bond constructs the programme's discourse as originating from the nerve centre and 'heart' of the British film industry, with the implication that it is offering cutting-edge, 'inside' news. Haigh and Bond frequently commented on industrial shifts within the film industry (such as the increase in independent production in Hollywood, the rising number of American films produced in Europe or the prevalence of particular generic trends) in what was intended to be a casual and knowledgeable manner. Many viewers were evidently impressed by this approach as Peter Haigh explained that they received 'hundreds of letters' each week 'asking for information of all kinds about films and film people past and present'.[32] Although he explained that many of them 'had to be passed to other sources' and that there just wasn't 'time to cope with them all', this suggests that viewers used the BBC as a kind of enquiry service (something long since offered by *Picturegoer*) which, on a very popular level, offers a further perspective on the links between the cinema programme and 'public service'.

The attempt to offer an 'inside' perspective on the film industry was all the more evident on *Film Fanfare*, and was taken to fascinating extremes in Peter Noble's gossip spot. In characteristic rapid-fire fashion, he would report on the films being produced, which stars were working where, and news on up-and-coming productions or plans. Noble always referred to the stars on very informal terms, claiming that they sent him personal telegrams, telephoned him, or called by his home. When Diana Dors was travelling to America in 1956, Noble read out a telegram to viewers: '"Having a wonderful crossing, love to *Film Fanfare*, and especially to you, Diana Dennis". That's the fabulous Diana Dors, thank you Di...'.[33] He then opened another telegram: 'Expect to see me in a week or two stop. ...Jimmy G', and explained, 'well Jimmy G is Stewart Granger, known to all his friends as Jimmy'. Noble gave the impression that he knew the stars on a personal level and offered a similar privileged knowledge with regard to film production with his up-to-the-minute schedule and newsworthy gossip. Noble's news generally centred on British stars, films and studios and from a contemporary perspective, it has an appealing insularity. It was a direct descendent of the shared world of British cinema constructed and circulated by *Picturegoer* - affectionately referred to by the magazine as 'British filmdom' - in which all the British studios are known by name. Marking its historical specificity, there is certainly a tangible sense of the 'local' in Noble's gossip spot in a way that differentiates it from the cinema programme today. As we shall see, the programmes also had a particular appetite for Hollywood cinema, although perhaps partly shaped by *Film Fanfare*'s links with ABPC, the ABC series did seem to devote a greater proportion of its time to British cinema. Clearly, however, the use of film actors as presenters was perhaps intended to blur the boundaries between the film industry and the programmes. Rather than simply reporting *on* British film culture, they offered themselves as an extension of the industry, and this was particularly so in the case of *Film Fanfare*.

This was then reinforced by the visual spaces which structured the programmes. In terms of the earlier series *Current Release*, the set was the site of a struggle between the BBC, the film industry and the audience, which in many ways reflected upon the instabilities of television's developing modes of address. It also reflected upon perceptions of how television should deal with and present the 'world' of the cinema. The first thing to note is that, like *Current Release*, the different spaces in the sets are fascinatingly ambiguous. *Film Fanfare*'s set was more spacious than *Picture Parade*'s - something exploited in the opening titles. Certain existing editions begin with a high-angle shot of an orchestra, staggered on platforms, as the words '*Film Fanfare*' appear in fancy scripture accompanied by stirring orchestral music (Figure 4).[34] The camera then tracks down to meet the orchestra at eye level and in several editions we see an enormous studio arc light on a platform which is directed at the musicians by a technician. It is then switched on with a loud 'click' as the orchestra leaps into action.[35] This not only gives the impression that we are witnessing part of the production process, but it also emphasises the *staging* of the programme, and the construction of the orchestra (with their suits and bow ties) as a spectacle. The opening of *Film Fanfare* seems to evoke the grandeur of the theatrical, rather than cinematic, in which a concert orchestra is brought to the home. In this respect, it is perhaps worth noting here that it is the ITV programme which in fact gives the impression of a rather middle-class formality. In comparison, the BBC's *Picture Parade* begins with the presenters aiming to give the impression that they are chatting 'casually' as the title credits roll down.

Other editions of *Film Fanfare* began in different sets and like *Current Release*, a central space was a desk area including a telephone and bookcase. This expansive space was again surrounded by a backdrop of large framed stills of film stars ranging from Audrey Hepburn, Shirley Eaton, Yvonne Furnaux and Terence Morgan, and these images were usually visible in the background of the shot (Figure 5). The desk was situated in a large, open-plan 'living room' area with a

Figure 4: Opening titles of Film Fanfare (1956, edition 15)

sideboard, coffee table, sofa, chairs and brightly patterned wallpaper which, in many ways, seems to reflect the 'lightweight' 'contemporary' design of modern furniture in the period (Hopkins, 1963, p.329). The desire for an uncluttered and thus more expansive space supports Lynn Spigel's discussion of the post-war American home and the promotion of 'the functionalist design principles of "easy living"' which eliminated walls in the central living spaces of the domestic sphere (1992, p.6). Home magazines spoke of the creation of a continuous space or an 'illusion of spaciousness' so that 'domestic interiors appeared not so much as private sanctions which excluded the outside world, but rather as infinite expanses which incorporated that world' (Spigel, 1992, p.7). As Spigel notes, television was not only a shaping factor in this functionalist reorganisation of space (an open-plan design means that it can be viewed from the kitchen, for example), but it also acted as its central metaphor: bringing 'another world' into the home with no necessity for movement on the part of the viewer (p.7). It is possible that *Film Fanfare*'s streamlined furniture, jazzy wallpaper and open-plan aesthetic were designed to reflect this new contemporary style space, and the values it was seen to convey (Figure 6). Yet it is difficult to see the set primarily as a consumer space, intended to solicit a domestic, consumer gaze, as it co-existed alongside other spaces which

Figure 5: Paul Carpenter in Film Fanfare (1956, edition 8)

attempted to simulate areas in a film studio. These were filled with the technological apparatus of filmmaking including cameras, arc lights, props and set scaffolding, and this technology (and working space) conflicted with the connotations of a domestic area. As with *Current Release,* it was unclear what *Film Fanfare*'s set, in its entirety, actually proposed to be. Was the lounge-room set and desk the reception area of a film studio?

Film Fanfare later moved from its original Sunday afternoon slot to Saturday evening, and it was at this point that the presenters all appeared in dinner jackets and bow ties - not an uncommon sight on 1950s television in many types of programming. Sylvia Peters, the post-war BBC television announcer, recalls the glamour of early television when discussing how the 'image, funnily enough, was incredibly glamorous. We had to change into full evening dress for the evening programming, [while] at the same time being rather matey and friends with the public'.[36] Peters describes here a certain tension between the need to address the audience in a 'friendly' and relaxed manner, yet the extremely formal and glamorous image she was required to project (and, of course, this was inflected by gender roles in different ways). While indicative of the more formal dress codes of

Figure 6: Peter Noble in the 'living-room' set (1956, edition 8)

the time, Peters' description of changing into evening dress for evening schedule also suggests something of the ways in which television sought to establish relations with its audience in the domestic sphere. Clearly drawing on very middle-class codes of conduct, it is assumed that if 'guests' are welcomed into the home in the evening, they wear evening attire. The development of the schedule around conceptions of family and domestic routine was central to the construction of the relations between television and the home, yet the discussion of costume here also suggests how television sought to build a *temporal* relationship with its audience in ways which were part of a conception of daily structure. That said, as an institutional practice the importance of costume here was not specific to television. Even the BBC's radio announcers had worn formal attire (surely a derivative of Reith's emphasis on setting impeccable social standards). Nevertheless, when seen on television it added an air of glamour to the proceedings, particularly in the context of the 'showbusiness' associations of weekend evening television. As explored in Chapter Five, the cinema programme revelled in the visibility of the cinema in the public sphere - stars arriving at crowded airports, the hustle and bustle of film premieres. The costume worn by all of the *Film Fanfare* presenters seems to retain the charge of this glamour, bringing it back into the more 'domestic' connotations of the programme's main set. In some ways this quite literally played out the power of television to bring the glamour of the cinema to the *home*.

Aside from the living room set, most of the other areas in *Film Fanfare* attempted to simulate a film studio. The most obvious use of this iconography was in Peter Noble's 'gossip spot'. Situated among arc lights, wires, set scaffolding and booms, he sat in a director's chair (Figure 7). In comparison with the lounge set, these areas were rather dimly lit with shady passages (almost like a backlot on a film noir set) and much of the apparatus looks as though it is being stored. This background is contrasted with the very bright light illuminating Noble as he sits in his dinner jacket and bow tie, poised in the director's chair. Various sets sought to give the impression that the presenters were behind the scenes in a film studio. Carpenter and Noble were often situated amongst arc lights, cables, booms and set scaffolding - mirroring the apparatus seen in the real footage of films in production (the construction of behind-the-scenes footage of filmmaking is explored in Chapter Six). These connections are particularly interesting given ABC Television's links with ABPC, and *Film Fanfare* was produced at Elstree studios, one of the busiest centres for British film production. However, when used in the programme, the studio equipment took on the status of a prop which was used to create the sense of a 'realistic' film studio. In this way, the set was designed to contribute to the impression that the programme literally occupied a space 'inside' the film industry.

John Fitzgerald's spot was called 'Movie Magazine' (borrowing the title from his earlier ATV show), and this was the section of the programme which previewed the films. Fitzgerald discussed the films in what was known as the viewing or 'preview' theatre - visually very similar to a small local cinema. With an 'Exit' sign was visible on the wall behind him, Fitzgerald sat at a desk in front of two rows of cinema seats (Figure 8). The excerpts appeared on the 'cinema screen' and the lighting was very subdued in order to simulate the construction of the cinema space. Not dissimilar to *Current Release*'s emphasis on the use of a projector to screen the clips, this scenario worked to differentiate cinema and television in terms of technologies and spaces of

Figure 7: Peter Noble in his director's chair in the film studio set (1956, edition 19)

Figure 8: John Fitzgerald in the 'viewing theatre' set (1956, edition 15)

exhibition. Despite the fact that they were soon to appear in increasing numbers on television, the use of the preview theatre perhaps suggests that the cinema is essentially the 'appropriate' place in which to enjoy film. This was of course integral to the commercial logic of the programme (particular where ABPC and ABC TV were concerned). On another level, however, the set also reflected back on the power of television to domesticate the consumption of the screen image. With Fitzgerald sat alone' in his small cinema, this evoked connotations of the increasingly exclusive and 'private' consumption of screen entertainment. At least from a retrospective point of view, it is also difficult not to connect this image to the wider economic reality of the relations between cinema and television at that time in so far as, sat in a dark and clearly *empty* cinema, Fitzgerald looked almost lonely. As discussed earlier, 1956 was the first year of significant cinema closures, although any reflection of the changing nature of the cinema's economic fortunes and social role is never acknowledged explicitly in *Film Fanfare*. The cinema programme effectively masked these changes on a number of different levels, as explored in relation to the construction of the behind-the-scenes footage (Chapter Six).

John Parsons was *Film Fanfare*'s 'roving reporter' and he also had his own particular space from which he introduced the footage of stars, premieres or films in production. Again constructed like an office, it included a desk, black telephone, filing cabinet and a large map of the world (Figure 9). Not only was he 'out and about' visiting press conferences or studio stages, but one of his main functions was to meet stars who travelled from abroad. As he explained, 'My job is to represent you and together we shall go out to airports, docks and stations and meet the stars as they arrive'.[37] The map, therefore, was perhaps intended to represent the 'global' reach of his task, bringing stars from faraway places into the domestic space of the home. In one edition, for example, he stands in front of his map and explains how it has 'been a truly international week with visitors from Hollywood, Rome and Paris...'[38]

In comparison, evidence suggests that the set of *Picture Parade* was less spacious. Panning from left to right (as the camera did at the start of each programme), the first space revealed was the interview set including a curved sofa and several large

Figure 9: John Parsons brings cinema back to the home: 'It has been a truly international week with visitors from Hollywood, Rome and Paris' (1956, edition 19)

bold pictures. Adjacent to this was the main set which included a long desk and black telephone, and this stood in front of three television monitors set into the wall. When not in use they displayed the words *'Picture Parade'*, matching the graphics of the opening titles (Figure 10). In front of the desk were two striped chairs with wooden arms which perhaps contrasted with the more simplistic, streamlined (and modern) design of the furniture in *Film Fanfare*. Next was the 'Hollywood set' which included a door labelled 'Library', although whether this was intended to refer to books or films is unclear, and its relations with Hollywood were ambiguous. As with *Current Release*, the BBC chose to include a secretary (named 'Olive'), and she occupied a desk towards the back of this set which sported a large white lamp, telephone, typewriter and papers. In front of Olive was the 'Hollywood desk' (normally used by Derek Bond) and although out of focus, Olive often remained visible in the background as she attended to her duties. (See figure 2)

The two remaining sets were known as the 'Theatre' and 'Foyer' sets. The Foyer set (as in cinema foyer) was used for entertaining the guests, and behind the sofa it boasted gold double doors attended by a uniformed commissionaire. The Theatre set was presumably intended to simulate a space in which films were viewed and in this respect was comparable to *Film Fanfare*. The viewing theatre

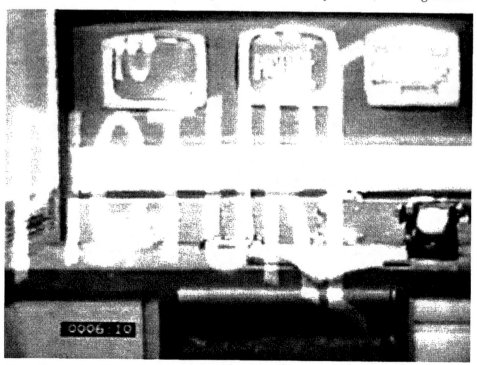

Figure 10: Opening titles of Picture Parade (1956, edition 18a)

was used by veteran British film director Maurice Elvey for discussion of historical aspects of film, and Elvey sat at a desk in front of a bold black-and-white diamond backcloth pattern (Figure 11). Outlets in the wall were intended to represent the channel for the projector beam and adjacent to Elvey were two red shiny seats, referred to in the scripts as 'cinema seats'. However, in comparison with *Film Fanfare*, the set did not so clearly resemble the interior of the viewer's local cinema - particularly given its bright electric lighting.

This discussion of the sets in both programmes indicates that there was again an attempt to draw on 'appropriate' cinematic settings on a variety of different levels. In fact, as with *Current Release*, the presenters would sometimes be 'caught' on the telephone while the programme was on air. In *Film Fanfare* John Parsons interrupted his address to the viewer to answer his ringing telephone, and received details of a star's arrival in the country: 'Shirley Jones... 9:45, London airport, flight 510, Monday...'[39] - such 'spontaneous' interludes again generating a sense of liveness in the construction of the cinematic world. In general, however, although *Picture Parade* and *Film Fanfare* in some ways explored the fictional aesthetic further by adding cinema foyers, viewing theatres and film studios to the repertoire, there in fact appeared to have been little tension over the sets. Neither the trade press

Figure 11: British film director Maurice Elvey in Picture Parade's 'viewing theatre' set (1956, edition 18a)

nor BBC Viewer Research indicate the vehement complaints directed at *Current Release*. This entire conception of a cinematic backdrop appeared to be better suited to the more glamorous and star-studded nature of the cinema programme at this time. However, while the aesthetic of the later programmes may have explored new spaces in which to weave their narrative of film culture, they were less inclined to incorporate the fictional aspects of *Current Release*. While moments such as answering the telephones were instances of fictionalisation which again required the viewer to suspend their disbelief, they were rare. Unlike *Current Release*, in which the sets originally structured Fitzgerald's role and his address to the audience, in the later programmes the sets really functioned as more of a backdrop - a visual context - in which the presenters moved. (Unlike the earlier secretary 'Susan', 'Olive' never spoke). Handing his coat to the commissionaire and striding through the foyer doors with a flourish, Peter Haigh's entrance at the start of *Picture Parade* clearly marked the commencement of the programme - its existence for the time the feature was on air - as well as the status of the studio space as a set. But while there were differences between *Picture Parade* and *Film Fanfare*, it would nevertheless appear that aspects of what were considered 'mistakes' in 1952 were later accepted as part of the *generic* iconography of the programmes.

The Generic Identity of the Cinema Programme

John Caughie describes the 'pervasiveness of *assumptions* of genre in writing and thinking about television, and ... the simultaneous difficulty in identifying where the theoretical grounding of these assumptions lie [original emphasis]' (1991b, p.127), while Laura Stempel-Mumford criticises the 'lack of an adequate theory of television genre as a whole' (1995, p.19). Although the issue of television's technological, aesthetic and formal difference from film was raised at an early stage (in John Ellis' *Visible Fictions* [1982] for example), such considerations were slow to pursue a comparison along generic lines. For some time, studies of television genre adopted models established by film theory (see Feuer, 1992) which, in view of the potential differences between the media, was not always satisfactory. One objection is that television is resistant as a medium to the type of categorisation genre definition requires. This is particularly so in the writing on contemporary television. As Stempel-Mumford explains:

> Television is seen ... as both the ultimate representative and primary purveyor of a postmodern sensibility, the site of self-reflexive mix of ahistorical pastiche... filled with programs [sic] that refer mainly to other programs... Its fluid formats, argue postmodern theorists, have borders far too permeable to fix into anything that resembles the genres of the past

(1995, p.20).

The brief reference to 'genres of the past' is indicative of the often dehistoricising (and decontextualising) thrust of postmodern perspectives on television, particularly given that there is so little work on the relations between television and genre from a historical point of view. The enormous amount of critical writing on soap opera, situation comedy or news suggests that genre is clearly a very visible factor in the organisation of television programming (and in academic approaches to the medium), yet in terms of contemporary television at least, programme forms have become increasingly blurred.[40] In contrast to the approach outlined above, other critics (Fiske, 1987, Caughie, 1991b) have pointed to the centrality of genre in television. In arguing that genre seems 'fundamental' to the medium, Caughie suggests that:

> In its endless repetition, week in week out, and in its flight from the prestige one-off production towards the reliability of the serial or the packaged series, television seems to involve a process of categorisation more intense ... than that of the mass production of Hollywood cinema in its classical period

(1991b, p.127).

Yet as Feuer points out, television's appetite for this ongoing structure is a reminder that, in comparison with films, television programmes do not operate as discrete texts to the same extent - they are part of a continuous flow (p.157). Although hybridity and intertextuality are also central to film genres, this creates a context in which television genres exhibit a greater tendency to 'recombine *across generic lines* [original emphasis]', what Feuer calls 'horizontal recombination' (1992, p.158). This equally raises issues about the (re)construction of genres from television's past - how texts were shaped by the wider flow of the television schedule, and how the existence and traces of this intertextual framework can be re-imagined. For example, while *Film Fanfare* included the items mentioned above (the interviews, the premieres, the 'behind-the-scenes' footage and so forth) these were interspersed with music from the orchestra and singing and dancing in what might be perceived as an almost illogical succession. As the title of the programme suggests (while also punning on the word 'fan'), the orchestra was seen to be a key aspect of its appeal. In introducing the items the presenters would simply explain 'And now for some light entertainment we turn to music' or 'Let's start the show off with a song...'[41] In terms of the song performances, *Film Fanfare* used a variety of singers including several minor film actresses. Appearing in her contemporary, white 'New Look' dress with clipped waist and full skirt, the American Diana Decker was described as the 'resident singer' and, unlike the others, she appeared more than once (Figure 12). They made no attempt to disguise what we might now perceive as a lack of continuity, or a rather fragmented format. In this respect, it is possible that *Film Fanfare* needs to be placed within the context of variety, a very popular television genre at that time - and on ITV in particular.

Figure 12: Diana Decker sings with the orchestra on Film Fanfare (1956, edition 7)

Sunday Night at the London Palladium (ATV 1955-67, 1973-74) was the most popular variety programme on British television in the 1950s and was rarely out of the top ten.[42] Produced by ATV, the series was transmitted from the world's number-one variety theatre and as well as the main acts it included Tiller Girls, music hall comedy and a quiz show (Evans, 1995, p.505). Robert Silvey, the Head of BBC Audience Research, noted that ITV 'did variety better' than the BBC and that they devoted more time to the genre which 'was what the public most wanted from television' (1974, p.164). ATV's managing director was Val Parnell and he was formerly managing director of the world's largest chain of theatres, including the London Palladium (Paulu, 1961, p.63). ATV in particular had access to the top variety stars in show business (Thomas, 1977, p.164), and titles such as *Val Parnell's Startime, Saturday Showtime* and *Variety Startime* appeared frequently in the ITV schedules. *Film Fanfare* was not a variety programme, but it at times seemed to reference the existence of a *variety* aesthetic.

The presenter's formal attire reinforced these links with the notion of a Saturday 'show' and variety, and in between items Paul Carpenter would sometimes stand in front of a long curtain like an announcer on stage. This was not necessarily an approach innovated by ABC, as American television programmes had integrated film and variety for some time. The most famous example was *Toast of the Town* (1948-55) (re-named *The Ed Sullivan Show* [1955-71] after 1955) (Evans, 1995, p.592) which combined excerpts from current films and interviews with their stars with acts by comedians, singers and other types of entertainers. Although American variety programmes such as *The Jack Benny Show* were screened on British television at this time, the British magazine *Picture Post* discussed the *Ed Sullivan Show* as unfamiliar when it noted that 'Its format appeals to American viewers, who like a melange of song, dance and fun',[43] and film was part of this hybrid text. What is important is that this 'melange' also recalls *Film Fanfare*, and the advent of ITV was indeed associated with the 'Americanisation' of British broadcasting on a number of different levels. This was usually with respect to the introduction of television advertising or the increased importation of American shows. If *Film Fanfare* was somehow 'Americanised', this was a more intangible and elusive link. As indicated in the discussion of ABPC, Warner Bros had held a one-quarter share in the company since 1941, and these relations may have been a significant factor here. Although it remains unclear how much influence they had on the decision-making at ABPC (and especially ABC TV), Warner Bros representatives would have been familiar with this variety format and *The Ed Sullivan Show* was renowned for its popularity.[44]

Equally, *Film Fanfare* also took a particular interest in different forms of quizzes. With Paul Carpenter acting as question master, *Film Fanfare* included a panel quiz in which two teams of three stars - usually British - were asked questions on film trivia sent in by viewers. A further element here was MacDonald Hobley's regular 'quiz' wheel sessions where viewers had to identify stars by dialogue excerpts from films, while there was also a chance to beat Peter Noble's knowledge of film trivia concerning film titles, dates and stars. Evidence also suggests that viewers were later invited to take part in the quiz themselves.[45] In fact, in August 1956, ABC announced plans to reorganise *Film Fanfare* almost entirely around a quiz format, re-naming it *Win Your Way to Hollywood* (in which viewers were offered the chance to do just that). This ultimately caused quite a controversy with the film industry which objected to the change (partly because it would reduce the time devoted to coverage of films), and the plans were later dropped. But it may well be significant to note here that *Film Fanfare*'s interest in such items was perhaps also shaped by the fact that quiz and game shows were central to the early schedules of ITV, and the building of its mass audience (Whannel, 1994, p. 183). Although the BBC had enjoyed success with panel games such as *What's My Line?*, it was ITV that were most clearly identified with the genre of quiz and game shows, which drew both large audiences and criticism about their populist and 'exploitative' nature (Whannel, 1994, p.182). Certainly, quiz items had been integral to the film magazine programmes on radio. Hobley's quiz section in *Film Fanfare* is the clearest reminder of this given that - focusing on dialogue excerpts with the camera pointed toward a rather cumbersome tape recorder - the item was certainly not very 'visual'. However, the marked penchant for quiz culture in *Film Fanfare* (and particularly the bid to transform the show into a quiz show itself), can again perhaps be linked to elements of channel identity and the surrounding programme culture in which it appeared.

These differences between differences *Picture Parade* and *Film Fanfare* do not so much undermine the claim of a generic status for the cinema programme, as they underline the complexity of this conception and of approaches to television genre more generally. This is particularly so from a historical point of view. *Picture Parade* and *Film Fanfare* retained core generic elements such as the (uncritical) film preview, coverage of stars and premieres and behind-the-scenes footage of filmmaking, and these items were presented within the visual context of a cinematic world. These contexts, while by no means identical, are linked by the generic iconography of the cinema programme. The use of preview theatres, cinema foyers, film studios and offices functioned to reflect and reinforce the programmes' populist approach to the cinema, and their attempt to construct a rhetorical closeness with the film industry - to convey the impression that their discourse originated from 'inside' this sphere. The differences between the programmes do not arise from the way in which they construct these core elements, but rather in the material which surrounds them. In the case of *Film Fanfare* this surrounding material is shaped by its relations with variety and quiz shows, which were not only wider elements of television flow, but were genres particularly popular on, and associated with, ITV. This in turn relates to the institutional basis of commercial television, and its economic imperative to broadcast more of the most popular genres to large audiences. This may well suggest, then, that one of the problems involved in researching the history of television and in focusing on particular series is that the texts can become isolated from their surrounding programme culture. This also indicates something of the complexity of approaching the differences between BBC and ITV which, in the context of a particular genre, are shaped by a range of different factors which cannot necessarily be reduced to rather homogeneous contrasts between presentational style, content and modes of address.

But the concept of a generic conception of the early cinema programme was also a site of discursive struggle the different parties involved. Caughie argues that scholars evince a tendency to adopt 'given institutional programme categories' (such as soap opera and sit-com, and this also emphasises how studies of television genre have more often tended to focus on fictional texts) (1991b, p.128). Yet of particular importance for historical analysis is that categories are not simply 'given': at some point they are *formed*. Steve Neale's 'Question of Genre' (1990) made an important intervention in approaches to the historical study of genres, while simultaneously shifting the focus of debate to encompass the intertextual construction of generic forms. Neale's emphasis on the intertextual and discursive construction of genre is not specific to film, and is useful in investigating the cinema programme. According to Neale, 'Industrial and journalistic labels ... offer virtually the only available evidence for a historical study of ... the way in which ... films [are] generically perceived at any point in time' (1990, p.52). This is part of the film's 'intertextual relay' which encompasses discourses of publicity, promotion and reception (Neale, 1990, p.49). Neale was interested in exploring the possible differences between, on the one hand, theoretical definitions of genre, and on the other, institutional and industrial uses of the term, locating a historical understanding of genre in the latter. In analysing the emergence of a generic identity for the cinema programme, the place to start is with the concept of the generic label.

While the New Historicism in genre criticism has opened up a division between industrial and critical categories, tensions can also occur *within* institutional, industrial and historical definitions. With *Current Release* there was a discursive struggle over how to classify the new programme. While BBC Viewer Research Bulletins described the programme as a form of 'Documentary', the film industry were inclined to use the term 'TV trailer' - a phrase also adopted by the press in conjunction with the pejorative use of the term 'TV plug'. The trade press also resisted the term 'film magazine' as they considered it inappropriate for a programme primarily conceived to focus on the latest release - and it had previously been used to describe the wider approach of film programmes on radio. It was not until 1956 when the cinema programme was scheduled on a weekly basis that a common terminology, that of the film magazine, was adopted in the trade press and the television viewing guides (and is still used in the television schedules today). It was not that the film industry had changed their conception of these features, or their suspicion of what they perceived to be 'extraneous', magazine material, but there was a demand for some kind of common terminology as more and more film review programmes proliferated on television. If texts are categorised generically at a discursive or intertextual level, this must have some relation to their content and form. A key debate in genre studies has been the problems of producing exclusive definitions which, in order to construct the generic corpus, privilege sameness over difference, or ignore the importance of generic hybridity (see Neale, 1990). It is true that *Picture Parade* and *Film Fanfare* adopted different formats but they were nevertheless perceived to perform a similar function and their content (previews, premieres, star interviews) was certainly very similar.

Current Release raised the concept of verisimilitude, and issues of genre and verisimilitude are inextricably linked (Neale, 1990). Caughie has suggested that the concept is more relevant to analysing fictional genres than television's 'range of non-diegetic... and often speech-led' forms of programming (1991b, p.99). The early cinema programme, however, blurred the boundaries between the diegetic and the non-diegetic, and the concept of verisimilitude is productive in highlighting how this process actually works. Neale suggests that, in interacting with generic codes, verisimilitude circulates as a system of expectation between text and audience, and generic verisimilitude basically refers to the 'rules' of the genre (1990, p.46). From this perspective, what we are willing to accept as plausible is not established in relation to 'reality', but as a function of already existing texts (Aumont, 1990, p.117) (see also Todorov, 1981, p.18). This thus has little to do with the text's relations with reality and cultural verisimilitude, but everything to do with the repetition of generic codes and their acceptance by the audience. In relation to *Current Release* it can suggested that part of the tension surrounding the programme was that it had no established generic codes upon which to draw, and thus no tacit agreement between text and audience. For a range of institutional, cultural and technological reasons, formats used on radio were of limited use when it came to producing a televised cinema programme. This does not necessarily explain why such codes became more acceptable in 1956, but objections were certainly less pronounced. Repeated across what was increasingly perceived as a *type* of programme it appeared that aspects of 'fictionalisation' became plausible and acceptable as an *effect* of the genre - not because they were seen as 'realistic'.

'Organised grovelling and puerility?': Still searching for 'Film Criticism'

The shift toward a more glamorous aesthetic (and the attempts to construct a discursive closeness with the film industry) only served to fuel the hostile responses from the British press. With feature films still in short supply, at least when *Picture Parade* and *Film Fanfare* first began, they continued to dismiss the cinema programme as a plug or bargaining tool with which television hoped to make the film industry more co-operative. The *Daily Mail*, for example, referred to *Picture Parade* as a 'Wardour Street handout'[46] while the *Sunday Times* considered it 'organised grovelling and puerility'.[47] The *News Chronicle* labelled the programme as 'simply fan-club stuff, television fawning humbly at the knees of mother cinema'[48] and implicit in all of these comments is the belief that the BBC is shamelessly begging the film industry to relinquish its feature films. Referring to both *Picture Parade* and *Film Fanfare*, the cynical *Punch* dubbed them a form of 'sponsored screentime' and argued that they encouraged 'an indolent acceptance of screen entertainment [and]... undermine[d] the role of the legitimate critic'.[49] As seen in relation to *Current Release*, the reception of the cinema programme in the 1950s was shaped by the concerns over television advertising at that time - particularly around the introduction of ITV and the regulations concerning the separation of programme and advertising material. In fact, *Punch*'s belief that they represented a form of 'sponsored screentime' takes on new inflections when situated within the context of these wider institutional and cultural concerns about the future of British television. This comment is precisely indicative of the confusion and ambiguity which still surrounded the term 'sponsored' broadcasting, and the hostilities toward advertising which underpinned it.

The cinema programme was still characterised by an ambiguous relationship with film criticism. Highlighting the significance of *Current Release* in setting out certain institutional and textual parameters in this respect, it was automatically accepted by all parties involved that *Picture Parade* and *Film Fanfare* would follow its precedent. But both the BBC and ITV adopted a defensive approach in which they incorporated an awareness of these criticisms at a textual and intertextual level. One way in which they dealt with the problem was to emphasise that the presenters were not 'actual' critics - and thus the intention of the programmes was *not* to offer a critical *appraisal* of the films. In the *Radio Times* Peter Haigh explained that:

> I'm not an actual critic, nor is Derek Bond. I hold critics... in high regard, but they will be the first to tell you that very few people in the major cinemagoing areas in the Midlands and the North ever read criticisms anyway. They have other means of assessing their entertainment.[50]

Haigh's comments, drawing on the working-class associations of the 'North', also constructed *Picture Parade* as the 'people's' cinema programme. Indeed, the BBC's initial desire that the presenter should be a 'representative of the ordinary filmgoer' takes on new inflections once faced with the competition from ITV and the need to attract a mass audience. The attempt to distance the genre from professional criticism also permeated the programmes themselves. Before it merged with *Film Fanfare*, John Fitzgerald explained to the viewers of *Movie Magazine* that he was 'not here to tell you what you should and shouldn't see, my purpose is to tell you a little of each picture, show you an excerpt... and I think you'll find that this is sufficient for you to make up your own mind...'[51]

In *Current Release*, the BBC aimed to incorporate an educational and informative approach to film, not only as shaped by an ethos of public service, but to balance the promotional implications of the programme. To be sure, *Current Release* primarily functioned at a popular level with the amiable Fitzgerald offering a cheerful survey of the latest release. Yet shaped by the need to justify its place in the television service (and the BBC's perception of what that service should provide), *Current Release* aimed to educate and inform, as well as entertain. This is an approach less apparent in the later programmes, and certainly in the general film review programme today. With competition between channels the need to attract, entertain and retain the audience exerted certain pressures on the content, form and address of the cinema programme. That said, *Picture Parade* occasionally considered the wider contexts of film culture beyond the focus on the stars or the latest release - something which differentiated it from *Film Fanfare*. For example,

Picture Parade alerted the viewer to film exhibitions and on one occasion, covered a festival of industrial films.[52] In an attempt to expand its coverage of film culture, *Picture Parade* occasionally previewed Asian, Russian, Japanese, Chinese and Yugoslavian films, sometimes in conjunction with screenings at the National Film Theatre (NFT) in London and other specialist cinemas. Maurice Elvey's spot on historical film culture was also outside the scope of a contemporary cinema review, although it was not overly didactic in tone. It is significant to note here that Elvey was the most prolific director in the history of British cinema. Between 1912 and 1957 he directed a phenomenal amount of films ranging across a variety of different genres. Although rarely a big critical success, his films were very popular with the public, particularly in the late 1920s and early 1930s (Quinlan, 1983, p.88). Situated in the 'viewing theatre' with its shiny red cinema seats, Elvey organised his spot around a particular theme or genre. Often focusing on one of Elvey's own films (sometimes from the silent period), it occasionally incorporated a directorial point of view, but it was usually just a brief illustrated talk with the aid of excerpts and stills. For example, in one edition Elvey discussed films about the legendary highwayman, Dick Turpin, including his own picture, *The Ride to York* (1917), largely considering the different screen versions of his character to emerge.

Although it was a focus on stars, premieres and 'behind-the-scenes' footage which predominated, the structure of *Picture Parade* was perhaps symptomatic of the BBC's uncertainty over a more popular programming policy at this time of institutional and cultural change the for television. From the BBC's point of view, Elvey's spot or the promotion of programme at the NFT perhaps functioned to temper the 'excess' of the programme in all its 'showbiz' glory. With excerpts from silent films juxtaposed with a review of *Rock Around the Clock* (Fred F. Sears, 1956), or an encouragement to sample a Russian film followed by a glittering star-studded premiere, *Picture Parade* literally exemplifies the BBC's programming and scheduling policy at this time: the mix of both mainstream and minority fare. In introducing the item on industrial films Peter Haigh began with the qualifier: 'Now just in case you think this sounds dull... we'd like to prove how first-rate [industrial films] can be...',[53] which suggests a hesitation, an uncertainty. In attempting to pre-empt their response, the BBC no longer seems as confident about conceptualising the needs and interests of its audience, particularly when, at the flick of a switch, they could tune into an alternative programme on ITV.

As 1956 got under way, film culture increasingly permeated British television, and intersected with many areas of film culture. In terms of the study of film, this is precisely where the critical and theoretical significance of the programmes is most apparent. The following two chapters explore in detail film stars on television, coverage of film premieres, and television's look 'behind-the-scenes' of film production.

Notes

1. Memo on 'ITV schedule', 20 September 1955, T36/3.

2. *TV Times*, 11 June, 1956, p.21.

3. *TV Times*. 25 September, 1955, p.22.

4. Data taken from 'Top 20' tables in *40 Years of British Television* (1992) by Jane Harbord and Jeff Wright, p.8.

5. *TV Times*, 19 February, 1956, p.23.

6. *TV Times*, 25 November, 1955, p.30.

7. *Radio Times*, 4 October, 1956, p.44.

8. *Radio Times*, 7 December, 1956, p.48.

9. 9 April 1956. 'Film Magazine and film excerpts in Television programmes'. T16/72/5.

10. Cecil Madden, 2 January 1956. T6/240/1.

11. Madden to McGivern, 2 Jan 1956, T6/240/1.

12. 9 April, 1956. T16/72/5.

13. *Today's Cinema*, 2 February, 1956, p.3.

14. See the article 'The studio that isn't scared of TV' in *Picturegoer*, 29 March, 1958, p.35.

15. *Daily Film Renter*, 4 January, 1956, p.1.

16. 1 April, 1953. Minutes of a meeting between the Better Business Committee and the BBC. T6/103/3.

17. *Daily Express*, 17 April, 1956. BBC press cuttings box, P662.

18. *Daily Film Renter*, 20 January, 1956, p.1.

19. *Daily Film Renter*, 15 June, 1957, p.4.

20. *Daily Telegraph*, 21 January, 2001.

21. *Picturegoer*, 15 September, 1956, p.20.

22. 'Blimey - Him Again', *Picturegoer*, 15 September, 1956, p.20.

23 'TV Ends Those Six Lean Years', *Picturegoer*, 8 December, 1957, p.8.

24. *Picturegoer*, 8 December, 1957, p.8.

25. *Picturegoer*, 8 December, 1957, p.8.

26. As above.

27. *International Television Almanac of 1963*, p.112.

28. *The Sun*, 4 April, 1973. BFI microfiche on Hobley.

29. Peter Noble obituary, *Screen International*, 22 August, 1997, p.4.

30. *International Television Almanac of 1963*, p.8.

31. *Picture Parade* script, 51a.

32. '*Picture Parade* and *Movie-Go-Round*', *Radio Times*, 8 February 1957, p.5.

33. *Film Fanfare* edition no.19. June 1956.

34. See, for example, *Film Fanfare* no.15, May 1956.

35. The most striking example is *Film Fanfare* no.7, April 1956.

36. *A Night in with the Girls*, Part I (tx 13 March 1997, BBC2).

37. *Film Fanfare* no.9.

38. *Film Fanfare* no.10, April, 1956.

39. *Film Fanfare* no.9.

40. One of the most prominent and current sites of debate in this respect is Reality TV. See Jon Dovey's *Freakshow: First Person Media and Factual Television* (2000) and Su Holmes and Deborah Jermyn (eds.), *Understanding Reality TV* (2004).

41. *Film Fanfare* nos. 15 and 10.

42. See viewing figures in Harbord and Wright (1995), pp.132-137.

43. *Picture Post*, 1 August, 1953, p.9.

44. That the programme had a well-known reputation is also suggested by the fact that, on certain occasions, *Picture Parade* would use interviews with American stars or directors that were originally from *The Ed Sullivan Show*.

45. The listings in the *TV Times* promise that 'filmgoers will be introduced by Macdonald Hobley and will take part in a film quiz judged by Peter Noble' (29 September, 1956, p.25).

46. *Daily Mail*, 8 December, 1959. BFI microfiche on *Picture Parade*.

47. *Sunday Times*, 20 November, 1958. BFI microfiche on *Picture Parade*.

48. *News Chronicle*, 8 December, 1956.

49. *Punch*, 22 August, 1956. BBC press cuttings box, PP62.

50. '*Picture Parade* and *Movie-Go-Round*', *Radio Times*, 8 February, 1957, p.5.

51. *Movie Magazine*, no.1.

52. *Picture Parade* script no.23b, October 1957.

53. *Picture Parade* script no.23b. October, 1957.

Chapter Five

'As They Really Are, and in Close-Up': Film Stars on 1950s Television

Part two of the book moves on from examining the development of the cinema programme as an institutional, textual and cultural form to focus more on its relationship with the parallel world of film culture. From the perspective of the small screen, television actively contributed to the daily rhythms and experience of the cinema in the late 1950s, in some ways quite literally capturing what this culture *looked like*. The critical, as well as historical importance of this perspective makes it possible to explore how knowledge of the cinema programme connects with, and intervenes in, certain critical debates in film studies, and this chapter is specifically concerned with the relations between film stardom and television.

After the cessation of *Current Release* in 1953, when the cinema programme re-emerged in 1956 it was immediately evident that film stars were to be an integral aspect of its form. There had been several editions of the earlier *Current Release* when no stars (or directors) appeared, with the film clips functioning as the structuring force. In the intervening years, film stars often appeared on general BBC magazine programmes, and this continued when ITV emerged. General interest or magazine series such as *In Town Tonight* (BBC), *This Week* (ITV), *Tonight* (BBC) and various 'women's' programmes, included interviews on a regular basis, and the star biography features such as *Portrait of a Star* (ATV), *Filmtime* (BBC) and *Close-Up* (AR) sometimes included interviews combined with an overview of the star's career. But in the cinema programme from 1956, it was almost a case of the excerpts taking second place to the interviews, and it was clear that televised coverage of the film star was expanding rapidly.

Certain industrial and cultural factors appear to have encouraged this shift. Firstly, it was the result of improved relations between film and television. Although successful, *Current Release* was an experimental venture and many film companies were certainly more 'suspicious' of television, as the producer clearly found out when he tried to secure star appearances. By 1956 regulations for aiding co-operation over the cinema programme had been established (how the clips were to be distributed, how requests for excerpts were to be made), and this also facilitated the arrangement of star interviews. While the American majors' concerns over the live (and hence 'unpredictable') nature of the television appearances continued, by 1957 *Picturegoer* was able to comment that in relation to Hollywood that 'the "no-appearance-on-TV" contract clause is practically a thing of the past'.[1]

This was also related to a change in stars' contracts themselves, at least where Hollywood was concerned, something which raises the issue of difference between the British and American contexts. Along with the Hollywood studio system, the star system underwent reorganisation in the 1950s. With the decline in production, engaging stars in long-term contracts ceased to be cost-effective for the studios. In 1947, 742 actors were under contract, but by 1956 this had dropped to 229 (Balio, 1976, cited in Macnab, 2000, p.182). Performers were increasingly hired for individual films in which they were a one of a number of 'assets' or materials (e.g. director, producer, script, star) brought together as a 'package' for a particular production. This arguably created an environment in which stars could have greater control over their careers, and in organising production around single film projects the industrial value of the star (as a guarantee against loss) increased, enabling them to demand higher salaries (Macdonald, 1998, p.196). Certain stars set up their own production companies and became producers (which also had the advantage of saving on tax charges). This context may have influenced the availability of Hollywood stars to appear on television at this time, since to a certain extent, they had greater control over their public appearances. BBC correspondence indicates that when it came to *Picture Parade*, the producer often contacted the star directly, although there were still regular examples of appearances being arranged through a particular film company.

For a range economic, aesthetic and cultural reasons, British cinema has never operated a 'star system' comparable to that of Hollywood (Babington, 2001). On an industrial level, while there were eight American studios dealing in star contracts, this compared to only two in Britain: the vertically integrated Rank Organisation and ABPC, and neither had the resources or 'or the expertise in star-building' of the Hollywood majors (Macnab, 2000, p.173). However, the contract system at 'The Rankery' in fact blossomed in the 1950s, with Rank signing over 50 stars by mid decade. As Macnab notes, 'Ironically, at the time that the Rank Organisation was expanding its star stable, the Hollywood contract system was breaking down' (2000, p.182). This is not to suggest that the British 'star system' was in better economic health. Hollywood stars continued to overshadow British stars in terms of profile and earnings (Ibid). Related to this, it is only more recently that the history of 'indigenous', British film stardom has been addressed (see Macnab, 2000, Babington, 2001). As Bruce Babington observes, dominant strands of star theory are almost 'wholly Hollywood-orientated' (2001, p.3) and there is a need to acknowledge considerable differences – whether in terms of economics, aesthetics or cultural meanings – in the function and circulation of British stars which resists a simple (and usually unfavourable) contrast with Hollywood.

Picture Parade and *Film Fanfare* featured a broad scope of British and American stars, ranging from the Hollywood superstar, popular British stars, to many lesser known performers and starlets. As discussed in Chapter Seven, influenced by certain shifts in film exhibition and the increasing consumption of 'Continental' films in Britain, various French and Italian stars also made regular appearances on the programmes at this time. The BBC's *Picture Parade* seemed to feature more of the 'big' American stars, at least when it came to personal interviews, and this is despite the fact that ITV paid higher fees for star appearances. This was in fact made public in a *Daily Sketch* article about Diana Dors which explained how she had refused to appear on *Picture Parade* due to the BBC's low fees. The headline announced: 'Dors says No to BBC', and Dors explained how ITV usually offered 125 guineas for an interview, over 100 more than the BBC.[2] Correspondence certainly indicates how the BBC's payment was understood to be more of a 'token' fee, and the producer always wrote to the star personally to express his thanks. Although impossible to list all of the hundreds of star appearances here, American stars seen on *Picture Parade* included Jack Lemmon, Gary Cooper, Robert Mitchum, Marlon Brando, Gregory Peck, Joan Crawford, Kim Novak, Grace Kelly, Jayne Mansfield and Gene Kelly; while popular British regulars included John Mills, Joan Collins, Jack Hawkins, Dirk Bogarde, Alec Guiness, Richard Todd, Kenneth More, Anna Neagle, Phyllis Calvert, Richard Attenborough and Belinda Lee. Perhaps due to its more direct institutional links with the British film industry, it seemed that ABC's *Film Fanfare* featured a greater proportion of British stars, and Hollywood names were more likely to be seen in filmed items at airports and premieres. Partly occasioned by their geographical *presence* in Britain, the coverage of the Hollywood superstars was certainly charged with more of an 'event' status - and two key case studies in my analysis later focus on Joan Crawford (*Picture Parade*) and Marilyn Monroe (*Film Fanfare*).

A further factor in the increase of film stars on television was the competition between channels. At the beginning of *Current Release*, John Fitzgerald would list the films to be previewed while, acknowledging the fact that stars were a key weapon in the competition for audiences, at the beginning of *Picture Parade* Peter Haigh would often immediately announce the guests who would appear: 'Hello everybody. This week our list of stars includes Henry Fonda, Yul Brynner, Ingrid Bergman, Steve Cochran, Ann Blythe, Ralph Richardson, and Margaret Leighton...'[3] In 1952 television's penetration of everyday life and public consciousness was more limited. To many stars, television was then an unknown quantity and hardly a medium on which to secure widespread visibility. The BBC (and its critics) also seemed ill-at-ease with the concept of television offering a 'showbiz' aesthetic at this time. However, by 1956, Haigh's introduction was evidently an attempt to 'hook' the viewer for the entirety of the programme, and discourage them from tuning in to an alternative on ITV.

While these factors contextualise the increasing visibility of film stars on 1950s British television, they do little to suggest how such appearances may have functioned in an aesthetic sense: how the stars performed on television, how it reshaped their cultural circulation and star image, and how it might have influenced the relations between viewer and star. Symptomatic of the divide between film and television studies, this is an area which remains largely unexplored. Although there is some work on the shift of Hollywood stars to American television in the 1950s, in terms of drama and variety shows for example (Mann, 1992, Negra, 2002, Desjardins, 2002), it is structured by assumptions based on a somewhat polarised conception of cinematic and televisual fame. For example, John Ellis' now canonical *Visible Fictions* (1982) represented one of the earliest attempts to compare the institutional and textual economies of film and television, as part of which, he considered their construction of fame. Ellis' widely known argument insists that while the cinema produces stars, television (as shaped by its rhetoric of 'intimacy', familiarity and domesticity), presents the 'personality' (p.106). Ellis equally drew attention to the different role of intertextual circulation here by arguing that, while in film stardom there is a dialogue between 'on' and 'off' screen persona which contributes to the performer's paradoxical construction, in television, the two often become more 'entangled' or blurred (with the implication that we are more likely to perceive the television performer as 'being themselves'). The movement of performers across media boundaries (whether in terms of roles in both film and TV fiction, the screening of feature films on television or film stars appearing in TV interviews) has always been more fluid than Ellis' argument suggests, and his comparison is somewhat generalised and decontextualised. Yet even where crossover is acknowledged it seems to accept this earlier model without question. P. David Marshall's *Celebrity and Power: Fame in Contemporary Culture* (1997), for example, considers the appearance of film stars on late-night (live) American talk shows, and in demonstrating this position, his argument is worth quoting at length:

> Film stars, in their live appearances on the programs [sic], break the narrative clo-
> sure of their film texts. Instead of the displaced time of the filmic text, they enter
> into the current time of live television... The studio audience and the home viewing
> audience are acknowledged and looked at directly. This acknowledgment of presence
> serves to reduce the aura constructed by the narrative film, where the film actor lives
> in a world quite separate from the film viewer...[It] decontextualizes the aura of the
> star and re-creates the possibility of the star establishing a more personal and famil-
> ial public personality... Indeed, his or her new non-narrative-centered discourse is
> that of conversation with the program's host; the style of discourse is itself heavily
> invested with the ordinary, the everyday, the familial

(1997, p.126).

Marshall's argument is presented as a universal conception, applicable to any period, place or star, but questions clearly remain as the importance of historical context here. The historical analysis offered by Mann (1992) (exploring appearances of film stars in 1950s TV variety shows), as well as Negra (2002) and Desjardins (2002) (both examining the impact of television in the late 1950s on the images of particular female stars), still seem to be permeated by these *accepted* conceptions of cinematic and televisual fame. Summarising Mann, Negra emphasises how this work has:

> drawn attention to the destabilizing effects of television on Hollywood stardom, the most acute effects perhaps felt by so-called 'auratic' stars, those whose image of glamorous distance from spectators worked well in the context of cinematic stardom but appeared strikingly out of place in the domestic context of television

(2002, p.108).

Despite the often fascinating dynamics of their analyses, the pervasive description of television 'recycling' Hollywood stars (the title of Mann's article), seems to tell us as much about existing prejudices concerning televisual fame as it does about the historical specificity of this context in the 1950s. Equally, there seems little acknowledgement that film star appearances on radio had been institutionalised for over 20 years by this time, surely prefiguring some of the shifts toward domestication and proximity which are foregrounded here. This in fact emphasises how much of this analysis of film stars on television anchors the shift irreducibly to its *visual* dimension - in this respect arguably shaped by the much broader assumption that television 'reduces' the 'cinematic' on a number of different levels. In comparison with radio, the addition of the visual is certainly important here, but what about a time when, to quote Marshall, television is not yet entirely invested with 'the ordinary, the everyday, the familial' (1997, p.126), but when its modes of address, insertion into everyday routines and indeed relations with the *cinema*, are still in a process of becoming? In short, what seems lacking here is the acknowledgement that there was a time when film stars appearing 'on' television was a new prospect for both stars and viewers.

This is not to deny any broader differences between cinematic and televisual fame, at either a contemporary or historical level. It is true that a developing distinction between the film star and the TV personality was already evident during the 1950s (see Medhurst, 1991). Although continually reporting the movement of performers across media boundaries, as well as devoting more and more space to the small screen, *Picturegoer* was at pains to distinguish the spheres of film and television in much the same way as the work above. Television was invariably constructed as an inferior mode of employment, both financially and in terms of prestige, and *Picturegoer* emphasised its investment in the film star when it insisted that 'the word "star" is being bandied around too much in relation to TV ... [It is] better to regard TV personalities as meteors... [they] may flash, but they don't last ... TV is not a midget cinema ... No, first and foremost it's a developer of fireside friends'.[4] Similar to Ellis' and Marshall's conceptions, this understanding of the TV personality is configured around notions of familiarity - 'friends' who are invited into the home by the 'fireside' - and the question then remains as to what happens when the film star is inserted *into* this flow. It is worth emphasising here that there is also clearly a difference between a film star appearing *on* television for an interview, and a star moving over to television to work in television drama, for example. At a conceptual level, in the TV interview, the film star retains the primacy of their connection to the cinema (they are they to talk about their work as a *film* star). Arguably, then, they are no more being 'recycled' than they are in a fan magazine or radio interview.

In conceptions of stardom, the references to 'aura' are derived from Walter Benjamin's famous article, 'The Work of Art in the Age of Mechanical Reproduction' (1936). Benjamin, of course, argued that the mass production characteristic of the culture industry destroys the uniqueness, autonomy and authenticity of the work of art, and for Benjamin, this was equated with a decay in the 'aura' of the art object (1973, p.215). But when this notion of 'aura' is invoked to discuss the appearance of film stars on television (or even film stars *per se*), its use is not straightforward. Benjamin in fact considered the film to be one of the most powerful agents of reproduction (p.215), and in comparing the film performer with that of the theatre, suggested that the latter possessed an 'aura' unavailable to the film star. The actor in the theatre was present in the same space as the audience so a more 'personal contact' was made, and they could adjust their performance in relation to that audience (p.222). This was in contrast to the film star as 'the camera is substituted for the public. Consequently, the aura that envelops the actor vanishes, and with it the aura of the figure he portrays' (p.223). Given that Benjamin did not consider the film star to be 'auratic' in the first place, television is arguably not really changing anything here. It may well be 'reproducing' the image of the film star still further, but this is again just as true of other intertexts such as fan magazines. Perhaps when it comes to film stars on television, there is an implicit acknowledgement of Benjamin's suggestion that 'the social bases [sic] of the contemporary decay of the aura [is]':

> the desire... to bring things 'closer'.... which is just as ardent as the bent toward over-coming the uniqueness of every reality by accepting its reproduction. Every day the urge grows stronger to get hold of an object at very close range by way of its likeness, its reproduction

> *(1978, p.216).*

The notion of television bringing the star 'closer' to the audience (which then creates a decay in aura) is implicit in the quote from Marshall above. Here, as central to television's rhetoric of liveness and immediacy, this is particularly related to notions of direct address. But this clearly raises certain questions. The glamour photo or poster, for example, also reduce the size of the film star, employ a form of direct address, and bring the celebrity into closer contact with the audience. But the still image has a central place in the iconic construction of the star, so why should it be seen as reaffirming their glamorous and 'exceptional' nature, while television causes a depletion of this aesthetic? The conception of 'aura' is also necessarily a matter of perception (it cannot, after all, be 'measured'), and this is all the more complex when it comes to historical reception which, in this case, we haven't experienced first-hand.

The Star Interview: Distant Ideal or 'Fireside Friend'?

As central to the communicative discourse of broadcast talk (Scannell, 1991, p.1), conceptions of the TV personality stress the importance of an informal, relaxed rhetoric, a mode of address which (following radio), was pursued by television during the 1950s. It is difficult to convey from a contemporary perspective just how incredibly apparent this is in much 1950s television programming. For example, the *Current Release* scripts indicated that presenter John Fitzgerald was sometimes to 'straddle' a chair (sitting on it back-to-front) when addressing the audience, or that he was to place himself on the arm of the 'Chesterfield suite' in a 'leisurely' manner. This aim to create a casual mode of address was clearly cultivated by the later presenters, particularly when interacting with the stars. In one edition of *Film Fanfare*, Macdonald Hobley 'casually' straddled a chair when interviewing the British actress Jean Kent - creating a rather odd shot composition in which he dominates the foreground of the frame as he leans towards Kent on the sofa. The aim for a casual, relaxed atmosphere was reinforced by the construction of a physical closeness between presenter and star, and illustrative here is Hobley's interview with the British actor Terence Morgan who, although appearing in a number of British films in the 1950s, is perhaps now best known for his role as the father of a deaf girl in *Mandy* (Alexander Mackendrick,1952). With barely an inch between them, Morgan and Hobley are positioned on a two-seater sofa, facing front, their literal closeness emphasised by the tight framing and static camera. Although complementing the programmes' wider bid to construct an informal and 'intimate' relation with the stars, there may have been rather more practical reasons for these strategies. In early television drama, for example, many conversations take place in frontal two-shots, so that the characters (when in conversation) are placed side by side, rather than face to face. In a live studio, it was difficult to cut between actors in the field/reverse field without getting the other camera in shot. As Caughie explains, instead 'of cutting into the scene the camera observes it from the front - like a viewer in the stalls' (Caughie, 2000, p.48). In the studio interview we would not expect the 'characters' to be faced directly opposite one another, but this may explain the preference for frontal two-shots and the reluctance to cut in, or to alternate between host and star. Strategies for shooting the television interview, whether live or on film, were still developing at this time and in the case of the cinema programme, it worked to emphasise the physical

Figure 13: MacDonald Hobley 'chats' to British film star Terence Morgan (1956, edition 7)

closeness between presenter and star, and thus reinforced aspects of 'intimacy' and 'informality'.

The interview with Terence Morgan seems paradigmatic of the ways in which *Film Fanfare* in particular adopted an approach of 'matey' familiarity with the stars, and Morgan (and other British stars) also fit more seamlessly into the very middle-class rhetoric of its address. It may also be the case that it is impossible to separate strategies for drawing the star into the communicative style of broadcasting from wider national inflections in film stardom. In terms of television's construction of the British stars, a comparison might be made with the British film fan magazine and its rhetoric of familiarity surrounding the British performers. But in discussing the construction of stardom and fan culture in *Picturegoer*, Macnab has observed that:

> Despite the enthusiastic flag waving on behalf of British stars, ... [it] cannot help but defer again and again to Hollywood. It reads like a small-town catalogue in which local products are dutifully listed alongside the much more alluring merchandise from the big city stores

(2000, p.174).

Yet it seems difficult to entirely agree with a reading of *Picturegoer* on these terms. First, the magazine's construction of Hollywood cinema is more ambivalent than Macnab suggests, and second, what is perhaps most apparent in its mediation of British stars is indeed the discourse of familiarity, affection and 'friendliness'. These were the players that made up the British cinema in *Picturegoer* known to its writers and readers as 'British filmdom' - permeating the weekly rhythms of images, faces, names, films and letters. Rather than representing an unappealing 'local' product in contrast with the more alluring American merchandise, this is also about British stars representing what Babington describes as 'reflections of the known and close at hand' (Babington, 2001, p.10). Although of course also dependent on the specificities of particular stars, it is from this perspective that they may have slipped more seamlessly into the rhetoric of 'informality' and familiarity in the context of television address.

Through the actions and posture of the host, there was an attempt to draw the star into this rhetoric, in keeping with television's developing mode of address. This was an attempt, perhaps, to shift them into the discursive (rather than literal) space of the TV personality. (The stars themselves responded to this in different ways, as the example of Joan Crawford will make clear). A key element in the construction of the informal atmosphere with Terence Morgan was the manner in which Hobley crossed to the sideboard to get a packet of cigarettes, offering one to his guest. Smoking frequently occurred on both the programmes at this time, particularly in *Film Fanfare*'s panel quiz. When associated with film stars, smoking was evidently seen to connote glamour and sophistication, and it was not shrouded in such publicised health concerns as it is today. The connections were also played out intertextually in so far as film stars appeared in advertisements, endorsing particular brands. In the 1950s editions of *Picture Post*, for example, British stars (such as Phyllis Calvert and Jack Hawkins) were seen in cigarette adverts which interwove the product with references to the star's current film.[5] The depiction of film stars smoking in the programmes may thus have seemed natural, expected even, and as well as constructing a sophisticated 'adult' image, it was intended to contribute to the informal, relaxed atmosphere outlined. On *Film Fanfare*, for example, the cigarettes would sometimes be offered at the end of the interview in order to give the impression that the 'chat' was to continue after the camera had cut away. In fact *Kine Weekly* commented that the presenters on both programmes 'tried to give the impression of being on intimate terms with [the stars]..., presumably aiming at making the viewers feel the same'.[6] Viewer Research reports on *Picture Parade* repeatedly recorded how Haigh and Bond were seen as 'delightfully informal' and there was apparently 'nothing but praise for their friendly way of conducting the interviews'.[7]

The programmes, then, attempted to draw the film star into an informal, relaxed rhetoric for the duration of the interview. Yet unlike the host, the stars did not

Figure 14: Peter Haigh interviews Joan Crawford (1956, edition 18a)

address the camera at all. While they were marked as being 'outside' of their fictional roles, as just 'being themselves', this absence of a direct address perhaps had as much in common with the film performance. As *Kine Weekly*'s comment on the interaction between host, star and viewer suggests, the presenters functioned to 'stand in' for the audience. They *mediated* between the viewer and the star, while the star themselves maintained a certain distance. However, according to Marshall, the television interview draws the film star into a form of 'non-narrative centered [sic] discourse' which, by implication, clearly separates it from film. Yet in the early cinema programme, simply by entering the sets the stars became part of a semi-fictional aesthetic - a' world of filmmaking, gossip, viewing theatres, glamorous cinema foyers and so forth. This offered less a 'non-narrative discourse' than another form of 'narrative'. The stars pretended that the office, foyer and so on were real, in much the same way as they did with a set in film. However, the star was of course still perceived to be operating 'outside' of their fictional roles, and this perhaps played with the ordinary/ extraordinary paradox so central to the construction of stardom (Dyer, 1998, Ellis, 1992). The audience were privy to them 'relaxing', having a cigarette, just being their 'ordinary' selves, while the interview also foregrounded their extraordinariness and aloofness.

Although stars mostly refrained from addressing the camera directly, the interviews were nevertheless played out in a very self-conscious manner which implicitly acknowledged the presence of the audience. As Scannell describes, the organisation of broadcast talk operates in the form of 'double articulation: it is a communicative interaction between those participating in the discussion, interview [or] gameshow... and, at the same time, is designed to be heard by absent audiences' (1991, p.1). This constant acknowledgement of an 'absent audience' seems to complicate, or exist in tension in with, the comparison made with the film performance - what John Ellis terms 'the sense of overlooking something which is not designed for the onlooker but *passively allows itself to be seen* [my emphasis]' (Ellis, 1992, p.99). There were also certain stars who, although not acknowledging the camera for the duration of the interview, would address the viewer at the beginning and the end in unison with the presenter. For example, this is particularly striking when Peter Noble talks to the young British actress Sally Ann Howes, then appearing in the comedy *The Admirable Crichton* (Lewis Gilbert, 1957).[8] Presenter and star are positioned in a two-shot in which we are partly viewing them from the side, and Sally Ann Howes immediately turns to the camera and utters a very polite and deliberate 'Good evening', then turning her attention to Noble's questions - at which point the viewer returns to the textual position of an unseen (but present) witness to the conversation. This careful acknowledgement of the viewer clearly marked the interview as a *performance*, constructed for the benefit of an onlooker, and from a contemporary perspective the interviews appear very staged. Sally Ann Howes' address to the audience accorded with the very polite and mannered approach pursued by television at this time, yet it also enabled the 'presence' of the viewer, a third party, to be visibly encoded in the text in a self-conscious manner. This is much more evident in this period and it carries the hallmark of a new technology, not only in terms of how the presenter and star are seen to perform, but in their acute awareness of being watched, and the fact that this is the sole reason for their interaction. In many ways, it is possible that this conflicted with the aim of presenting an informal and relaxed mode of address.

With regard to *Current Release*, the stars who did appear on the programme were self-consciously introduced to the audience, not simply because television was a new medium, but because many of the guests were virtually unknown. They were required to tell their 'story' to the viewer - to explain how they had trained at RADA and the films in which they had appeared. By 1956 the majority of the stars were well known yet this verbal structure was still apparent, and it is by no means redundant in current celebrity interviews. This structure, however, appeared to be inflected in different ways in the 1950s. It was almost as though because the stars were being presented through a *new* medium, they had to be introduced to a *new* audience. This is despite the fact that by 1956, when set ownership was more widespread, the television and cinema audience would have become indistinguishable. So while the discussion of stars was indicative of a certain 'knowingness', an utterly *shared* knowledge of British film culture (particularly in

Peter Noble's 'gossip' spot, for example), this could contrast with their actual physical appearance. This again may complicate claims of the star establishing a more familiar personality, and the attempts to draw them into the rhetoric of broadcast flow.

The close relationship established between presenter and star differed from interview to interview. In the Terence Morgan example, the relationship between presenter and star seems balanced and 'matey', and Hobley and Morgan appear to be on equal terms. As the example of Joan Crawford will suggest, the presenter could also express an utter deference toward the star, and it would seem that, although the precursor of radio should be acknowledged here, there was little precedent for the striking flattery which could characterise the early television interview. The tendency of the fan magazine to construct the stars as 'friends', as well as its often more critical edge, meant that a similar approach was unlikely to be found in *Picturegoer*. According to the editor of the magazine, Robert Ottoway, both *Picture Parade* and *Film Fanfare* presented the stars as uniformly '"fabulous" or "astonishing"', and he noted with regret how the 'interviews were crammed with flattering platitudes. There is never any hint that there are degrees of excellence...'[9] Suggesting an interesting crossover with the magazine, Ottoway was actually the scriptwriter for *Film Fanfare*. His comments appeared in an article in which he criticised the film industry for forcing the programmes to operate within the constraints of the 'no criticism' clause - restrictions which he suggested encouraged a particularly 'enthusiastic' attitude toward the stars. While a growing number of current affairs programmes (such as *This Week*) were developing more trenchant investigative techniques in other spheres (see Wegg-Prosser, 2002), the film industry increasingly complained that the interviews were too polite and lacked controversy - somewhat ironic given the discursive parameters in which they had forced the programmes to exist.

While comments which indicate how viewers enjoyed '"meeting" all the stars' suggest a genuine sincerity and pleasure on behalf of the audience, Viewer Research reports also offered more negative responses: 'One knows in advance that all film stars are going to say how happy they are and how much they enjoyed working with so and so'.[10] While this view (expressed two years after *Picture Parade* began) could simply be interpreted as indicative of the increasing conventionalisation of, and familiarity with, the television interview, it also indicates the potential lack of precedent for its strikingly 'polite', sycophantic approach. These viewer complaints also tell us something about television's regulation of extratextual *knowledge* of the star, and its role in the discursive construction of the star image. Stars were generally asked about their future and current career plans, and television was drawing upon existing models in fan magazines and on radio. Ellis notes that fan magazines foreground the 'star-as-worker' more than other media (the press for example) (1992, p.96), and television continued this trend. Nevertheless, television seemed to occupy a particularly 'safe' ground in regulating extratextual knowledge of the star persona from the start. *Picturegoer* noted that the television interviews rarely searched for the 'facts': 'Nobody ever asks about money on television, it just isn't done. Nobody ever asks Sinatra if he and Ava are going to patch it up, or if Bacall is really the one'[11] and attributed this to the presence of the family audience (the very same audience which had of course historically been *Picturegoer*'s target). On radio, the BBC had maintained strict parameters around the star interviews, as illustrated by the description of material for *Woman's Hour* which should offer 'biographies of the film stars' focusing on 'their artistic careers rather than their private lives'.[12] Although much of radio's coverage of film was very popular in tone, 'gossip' about star's private lives was evidently seen as lowbrow and sensationalist, and certainly incompatible with the policies of public service. Speaking of television's *Picture Parade,* Haigh emphatically emphasised that although the interviews were not strictly rehearsed, 'The stars knew that personal or rude questions would not be asked ... that is not the way we work. A personal and private life should be treated as such'.[13] While the British fan magazines had not adopted the practice of indulging in lurid 'gossip' and generally presented a very conservative image of the star, by the late 1950s the *Picturegoer* began to slightly shift these boundaries. This was perhaps partly in response to the gradual loosening of film censorship, as well as the influence of American scandal magazines such as *Confidential* (see Desjardins, 2002), although the extent to which these magazines were available in Britain at this time is less clear. Nevertheless, in a bid to stem its declining readership, *Picturegoer* began to publish increasingly 'daring' headliners which promised the 'shocking truth' or some such revelation about various stars. As Mary Desjardins argues (in focusing more specifically on television's *This is Your Life* (1952)), both television and the scandal magazines were forms of 'new' media in the 1950s which worked to create knowledge about film stars. But in many ways, they worked in oppositional directions. The magazines were partly conceived in opposition to the growing popularity of television and perceptions of its conservative, familial values. One editor claimed, for example, that they offered the audience the type of star 'exposes' 'they can't get on television' (Desjardins, 2002, p.122). While the concept of the star image has historically been seen to pivot on contradictory meanings and incoherence (Dyer, 1979, Ellis, 1982), Desjardins suggests that, with the decline of the studio's controlling hand in the construction of the star image and the decentralisation of the authorship of star texts, such 'new' media contributed to a *multiplication* of contradictory discourses surrounding the star image. What we essentially have here in the 1950s are further sites which lay claim to the 'star-as-person' - their off-screen construction.

Television clearly needed to conceive a place for itself within the cultural circulation and mediation of film stardom, particularly if its discursive access to their off-screen persona was to be limited. In an early edition of *Film Fanfare* John

Parsons introduced the viewer to what they might expect from television's coverage of the film star:

> I think you would all agree that it could be interesting, and good entertainment, to go out and meet the famous film stars as they arrive in this country. Most of us have seen them in their big screen roles, but very few of us see them as they really are, as themselves, and in close-up.[14]

Not only does he promise to offer a glimpse of the 'real' person 'behind' the star, but this is anchored to what is presented as a specific attribute of television - its close-up. The close-up, of course, has more often been perceived as specific to film, particularly with reference to the representation of the star. Important in the growing public fascination with the film star (with its possibilities of close access and spectacularisation), the close-up is significantly seen as a privileged moment in disclosing the star's face: a 'moment of "access" to the star's private self' (Dyer, 1986, p.11). Yet the description of this moment as 'unmediated' is crucial, since according to *Film Fanfare*, this is what *television* was seeking to present. In a bid to construct the appeal of this new coverage, it is defined in opposition to the constraints and 'mediation' of the stars' film performance or 'big screen roles'.

Comprised of a range of texts, the construction of the star image constantly works to negotiate 'authenticity' (Dyer, 1987), to make the image 'something more ... more real - than an image' (Gamson, 1994, p.143). In terms of his bid to ask questions about the social and ideological function of stars, a crucial element of Dyer's work was to explore how stars articulated ideas about personhood - a site for the working through of discourses on the construction of identity, and individualism in particular.[15] In this respect, then working from a position influenced by Althusserian Marxism (see Lovell, 2003), Dyer argued that the textual and discursive construction of stardom could be seen to support the notion of individualism upon which capitalist society depends. In operating as part of this ideological 'fiction', he suggested that the continual insistence on 'authenticity' in the star image, or the perpetual attempt to lay claim to the 'real self', pivoted upon a particular model of personhood which endorsed the notion of a 'separable, coherent quality, located "inside" consciousness and variously termed "the self", "the soul", "the subject"...' (1986: 9). It insisted, he argued, upon the pursuit of 'some inner, private essential core' (Ibid: 12). This access is, of course, essentially illusory given that any 'intimacy' or knowledge permitted is entirely discursive, as for the audience, the star can only exist within representation (Marshall, 1997, p.90).

Television, then, is very much part of this matrix, and in many ways, fostered these claims still further. Within the framework of negotiating authenticity, it is entirely conventional for the intertextual mediation of the star to claim that it offers 'a privileged access to the real person of the star... an apparent or actual escape from the image that the [film industry] is trying to promote' (Dyer, 1998, p.61). Television rapidly became part of this textual regime. At a time when it was contributing to a decline in cinemagoing, it would be easy to see Parsons' introduction as expressing television's attempt to usurp the cinema's role as the primary place in which to experience the star, in 'close-up'. But intimacy is constructed here in relation to the particular aesthetic structures of television (and in ways which clearly differentiated it from coverage on radio). In the 1950s, television screens were often small, so the televisual close-up would not offer a much larger scale image of the human face than that which we encounter on a daily basis in everyday life, and we do not consider that as offering access to the 'private self'. Yet the construction of film stars specifically plays with notions of distance and aura, as Parsons deliberately invokes to particular effect. Specifically foregrounding the larger size of the cinema screen, he describes the televisual close-up as offering a more 'intimate' access to the star-as-person, exploiting what were seen as key characteristics of television at this time, intimacy and actuality. With a smaller image and closer relation between viewer and screen, television attempted to attract the early viewer by promising a more 'authentic' perspective on the film star which offered a new visual and discursive access to their 'persona' Presented as a privileged, visual perspective on the star, we are promised 'intimacy' here, but not necessarily (as critics later assume) in a way that is co-terminus with 'familiarity'.

Much of this seemed to draw not only on television's rhetoric of intimacy and actuality, but also the equation of 'liveness' with realism. As MGM's early concerns over stars appearing on *Current Release* suggest, this new context for the public visibility of the film star could also be seen in a negative light, particularly in terms of its implications for the star's image. Discussing the appearance of Hollywood stars on American television, but also making reference to the British context, a writer in *Picturegoer* insisted that:

> Film stars, forgivably egotistical, seem to be sucker-bait for personal appearances on TV - look at [the BBC's] ... In Town Tonight if you don't believe me. More often than not they appear to be just frightened amateurs. A long way from the closed, controlled seclusion of the film studio, they writhe nervously, held in enormous and frequently unflattering close-up. Yet they seem drawn to this business: they accept the challenge of TV without seeming to have understood its destructive implications.[16]

Here, rather than revealing the beauty of the face (as in the feature film), the televisual close-up is seen as 'unflattering' and implicitly intrusive - very different from the 'unveiling' of the star promised by *Film Fanfare*. The article also implies that the visual aesthetic of television (perhaps its different lighting, setting, use of make-up) can deglamourise the film star, again paralleling the arguments about a decline in the star's 'aura', distance and mystique. What is interesting is that according to *Picturegoer,* this deglamourising of the star works in dialogue with how they react to the TV camera, and it is the changed *temporality* of the appearance which is foregrounded here. The article 'TV Wins Over those Frightened Stars' (1955), for example, spoke of live television drama and explained how 'Stars who had no stage training and were not in the habit of acting more than two or three minutes at a time were scared silly of live TV and its split second demands'.[17] Similar concerns circulated around the interview or personal appearance which, having been less rehearsed, were perceived to be all the more 'dangerous' in this respect. What seems to be causing concern is the difficulty of film stars adapting their public profile and performance style to the new medium, tensions intriguingly displayed in Joan Crawford's first appearance on television, a debut she made on the BBC's *Picture Parade*.

'The Fabulous Visitor': Joan Crawford does *Picture Parade*

This case study, and the following example of Marilyn Monroe, are important in offering detailed analyses of how these star appearances were performed in ways which complicate the generalised assumptions made about television's impact on the circulation of the film star. At the same time, they also work to emphasise how television appearances by stars clearly intersected with the wider construction of their image at particular points in time.

Joan Crawford's appearance on *Picture Parade* was considered to be an important event, perhaps more than any other interview in the series.[18] Yet her high profile and fame make an analysis of her 'performance', and insertion into television's flow, all the more interesting. Crawford's appearance was linked to her visit to Britain to work on *The Story of Esther Costello* (David Miller, 1957), and in an unprecedented manner, the coverage spanned two editions of the programme. Haigh opened the first edition by explaining:

> We would like to introduce... a few of the many men and women who... bring you news of events and personalities through the columns of our national papers... Tonight we meet them on the job, and hear what they have to say about yet another fabulous visitor to Britain.[19]

The first edition featured the throngs of newspapers and photographers as Crawford arrived in the UK, emphasising the 'excitement' of her geographical proximity to UK viewers from the start. The following week Haigh introduced the programme by explaining 'We open *Picture Parade* with what might be termed "phase 2" in the visit of Joan Crawford to the country', and this edition showed footage of her visit which began with her appearance at the National Film Theatre. Prior to a screening of her academy award-winning picture *Mildred Pierce* (Michael Curtiz, 1945), she arrived to meet members of the BFI and answer their questions.

The scenes were transmitted on the monitor with Peter Haigh's commentary contextualising and clarifying the images seen:

..With her in the foyer are her husband, Alfred Steele, and twin daughters, Cathy and Cynthia... [Here is] Ernest Lindgren, the Curator of the National Film Archives... and a kiss for Jympson Harman, the doyen of the Film Critics... He assures Joan of a warm welcome awaiting her within the cinema... A photographer's last chance to get a picture before Joan Crawford went to meet members of the British Film Institute... And there she is, amid the applause of the crowd, taking her place on the rostrum ... Never lost for an answer, Miss Crawford was a vivacious subject for the questioners.

Haigh paces his description very carefully, and gradually it builds up to its narrative climax when Crawford rises 'amid the applause of the crowd'. This footage also builds up to the climax of Crawford's personal appearance which followed directly after. Despite her introduction, it is fair to suggest that the peak of Crawford's career had now passed. Like many Hollywood stars in the 1950s, Crawford was no longer contracted to a major studio (as she had been with MGM in the 1930s and Warner Bros in the 1940s). In 1952 Crawford had considerable success with her first independent film, *Sudden Fear* (David Miller, 1952) (starring with Jack Palance) and she set up her own production company (Considine, 1989, p.269). However, although she was still voted 'favourite actress' by American *Photoplay* in 1953, her next two films, *Torch Song* (George Walters, 1953) and *Johnny Guitar* (Nicholas Ray, 1954), flopped badly at the box office, and this was a troublesome time for the star. In 1955 Crawford entered her fourth and reputedly secure marriage with Alfred Steele (president of Pepsi-Cola), yet she had begun to drink heavily and was acutely aware of her declining career as an aging female star. With the decline of the studio system in the 1950s, the stars to a certain extent lost the protective barrier that existed between them and the press, and print media published increasingly sensational news stories about the stars' private lives. As Shaun Considine explains, 'The private lives of the stars, no matter how sacred, were no longer considered off limits... Crawford, "Saint Joan of the Fan Magazines", was one of the first to be burned at the tabloid stake' (1989, p.284). Just prior to the opening of *Johnny Guitar* in the US the *Los Angeles Mirror* asked to do a story on Crawford to accompany its release, part overview and part interview. This, however, turned into the series 'Joan Crawford - Queen or Tyrant?' in which family, stars and directors all denounced Crawford's domineering and erratic behaviour, and her off-screen behaviour - in the form of drunken incidents and sexual 'flings' - repeatedly received bad press reports (Robertson, 1996, p.100). As Pamela Robertson notes, Crawford's much publicised attack on Marilyn Monroe at the *Photoplay* awards ceremony in 1953 (at which she viciously criticised Monroe's 'provocative' attire), marked her as a 'bitter aging star', an image which the 'frighteningly self-referential' *Torch Song* reinforced (Ibid: 98-9). (*Torch Song* is a film in which Crawford plays bitter and lonely song and dance star who is utterly unpleasant to her colleagues but 'devoted' to her fans).

Although (when compared to its past), *Picturegoer* attempted to offer a slightly more revealing perspective on the stars by this time, this was not along the lines of the scandal magazines. This raises the issue of whether Crawford's construction in Britain differed from that of America, a context in which, although the British press could certainly be unkind to stars, their private lives were perhaps the subject of less extensive public knowledge. In fact, although it is of course a different type of publication to the press or scandal magazines, what is striking is the extent to which Crawford's construction in *Picturegoer* at the time of her visit offers a stark contrast to the image outlined. (Perhaps this is why she made her television debut in Britain, rather than America, where she felt she might be better received). These different constructions also provide possible frameworks of interpretation for the reception of her interview. In 1957 Crawford was mentioned in *Picturegoer*'s article 'Do you love a star who's forty?' which makes the sweeping complaint that 'The reluctance of the ... Joan Crawford generation to retire graciously from youthful roles - in short, to act their age - is partly responsible for the current drop in attendance', and insists that they should 'age' into 'character roles' rather than playing lovers in romantic stories (although some of Crawford's films in the 1950s - such as *Female on the Beach* (Joseph Pevney, 1955) and *Autumn Leaves* (Robert Aldrich, 1956) - combined this by placing her as the 'older' woman in relation to a younger love interest).[20] According to this particular article Crawford is no longer perceived as 'auratic' at all, largely because she has ceased to be romantically desirable, and the fact that she is at a point of change in her career is clearly noted here. An article designed to coincide with her visit to Britain in 1956 constructs her image differently. It notes that a number of her recent films (*Torch Song* [1953], *Johnny Guitar* [1954], *Female on the Beach* [1955], *Queen Bee* [Ranald MacDougall,1955]), were 'a long cry from the golden days of *Mildred Pierce*, ... and longer still from the days of *A Woman's Face* (George Cukor, 1941)' and argues that they have not been worthy of her talent.[21] However, the overwhelming emphasis of the article is on her status as a 'Hollywood legend ... one of Hollywood's Greats': '[She] has been, still is and always will be ... THE Joan Crawford... Stars have risen, stars have waned, but ... Crawford has remained, dazzling and indestructible'. The article attributes this to her 'intense dedication' and 'complete professionalism', well-known attributes of her star persona from the start of her career. There is an attempt here to return to the (ideological) stability of her traditional star image given that, rather than emphasising this time as a period of considerable change for Crawford, it glosses over these issues by stressing continuity.

This construction of Crawford as a Hollywood legend was partly shaped by perceptions of the decline of the star system at this time. *Picturegoer* offered a complex construction of Hollywood in the 1950s which alternated on a weekly basis between such headliners as 'Hollywood goes Bust!' 'Hollywood is Booming!', with television as the primary protagonist in this dramatic narrative. A 1956 article 'Can Hollywood Survive?' drew an evocative picture of Hollywood as a 'ghost town' with countless vacant studio lots. As part of this, it paid particular attention to the star system. While the reduction in star contracts was a reality, what is interesting is the assumption that if the 'star system has had its day', this was partly due to a dearth of new talent. The magazine enquired 'Where are the new generation of stars?... Hollywood's survival depends on your acceptance of a new crop of personalities'[22] and similar conceptions circulated in the trade press at this time. Joan Crawford, along with Clark Gable and Humphrey Bogart, is pictured as one of the 'Hollywood Greats', representative of the star system and the 'golden age' of yesteryear. Instead of foregrounding the decline of her career, Crawford's image is invoked to express a nostalgia for the past.

In contextualising Crawford's construction in Britain at this time, and in understanding how her television appearance both drew upon and contributed to this image, she was very clearly constructed as a Hollywood 'legend' by the programme, not least by Haigh's attitude toward her. Before introducing the star he solemnly explained (with great reverence and poise) how it was 'a very proud moment for *Picture Parade*' and at the end of the interview she and Heather Sears were presented with bouquets of flowers. Crawford's fame and prestige are also foregrounded through her interaction with Haigh. Rather than playing on his own emergent celebrity status, he expresses *utter* deference toward the star, and evidently feels that his meeting with her is an honour. The appearance of the young British actress Heather Sears during the interview, acts as a reminder that it is she who is actually the star of *Esther Costello* (as Crawford wistfully explains), and it was Sears who was awarded the title of top actress in the reader's poll in *Picturegoer* that year[23] (a list in which Crawford's name no longer featured). Haigh also asked Crawford to comment on the talent of 'today', which foregrounded her status as part of an earlier generation of stars, but the main aim here is to get Crawford to discuss the young stars she has helped to 'develop' ('Tell us about some of the young stars you have found, and helped along the way...?') which is ultimately channeled to reinforce, rather than diminish her legendary status.

Crawford certainly commanded attention. Seated on the sofa (to the right of the screen), she wore a black sequined dress and gold necklace which glittered under the studio lights and stood out starkly in television's modest aesthetic.(See figure 14) *Picturegoer* had praised Crawford's belief in being a 'star *off* screen as well as *on*'[24] and she significantly refers to the interview as 'this performance', almost as though she is describing a part in a film. This further raises questions about the ways in which film stars were assimilated into modes of broadcast address (or otherwise). For example, her performance is strongly 'melodramatic'. When Haigh asks her to explain the plot and theme of *Esther Costello* she does so very intensely. Clasping her hands and staring upwards (away from the camera), Crawford adopts a dramatic tone, almost as though she is reciting her lines: 'A woman ... goes back to Ireland, she left at six years of age... and this is the story of *many, many* women in the world. It doesn't have to be England, Ireland, Scotland, Wales...' At this point she appears to be totally absorbed in her performance and it is significant that *Punch* cynically observed a 'melodramatic' rhetoric in the interview on a different level when it claimed that it 'expected the censors to step in at any moment with a ruling about the duration of screen hugs, kisses and the laying on of hands'.[25] What is unfavourably received here is the 'excess' of the occasion, which is conveyed through bodily gesture, a key attribute of melodrama as a genre and performance style. With the intensity of her physical movement and taut body language this was a style with which Crawford was particularly associated and as a result, her interview pushes the comparison with her film performance to its extreme. This seems to come more naturally than the style required for television and clearly contrasts with its rhetoric of *understatement* and relaxed informality.

As Crawford performs this 'scene' in her glittering black dress, one of the reasons that she appears so striking is that her 'aura', her presence, is not so much 'depleted' by the aesthetic and textual context of the medium as it is *too excessive* for television at this stage. (Indeed, the first question she is actually asked in the interview is: 'Well, tell us something about "glamour"... What is your recipe for it?'). Her dress is in fact similar to that worn in her films. This is particularly so in the case of the black-sequined dresses worn in *Humoresque* (Jerry Wald, 1946) and *Possessed* (Curtis Bernhardt, 1947). In the 1930s, the glamorous nature of Crawford's costumes had been a recognised attribute of her star persona (Gomery and Allen, 1985, p.183) and this is replayed here to interesting effect. As her sparkling dress differentiates her from Peter Haigh and Alfred Steele (who appeared for part of the interview, holding hands with Crawford to talk about their 'perfectly ordinary life'), it recalls the way in which 'visually ... her heroines had overwhelmed their male counterparts' (Gomery and Allen, p.183). Yet the clash of image and performance codes here can be interpreted in different ways. Although the fitted, knee-length style of her dress is by no means dated, Crawford's appearance can be seen to evoke an older image of Hollywood glamour, and in offering an almost 'filmic' performance, it could imply that she is failing to adapt to the new television 'scene' and its implications for her profession. In the early 1950s, Crawford had in fact campaigned against film stars appearing on television, branding those who did as 'traitors' (Wayne, 1988, p.198), and while by 1956 she had evidently conceded to appear (and understood television's value as a promotional tool), it could be perceived that she continued to cling to her status as a traditional Hollywood film star. This supports the reading of Pamela Robertson (1996) who argues that extratextual knowledge about Crawford's private life, her association with 'artifice' (both in terms of appearance and behaviour), and her representation of an American femininity increasingly outmoded by the 1950s, all function to recode her image as 'camp' at this time. While it is often difficult to ascertain whether 'camp' is a retrospective reading (particularly when all of Crawford's classical Hollywood melodramas could be read this way today), Robertson argues that the 1950s is a period when Crawford's status as 'a "straight" star who appeals primarily to women and her status as a camp object briefly overlap' (1996, p.88). So while the visual and discursive construction of the interview accentuates Crawford's glamour and her exceptional identity as a star, how this was ultimately perceived is, of course, open to interpretation.

However, if Crawford's 'aura' can be seen to be too big for television, she is also acutely nervous in a way that conflicts with this - something crucial in understanding the live nature of her appearance. Although film stars had appeared on BBC radio for many years, whether in interviews, adaptations or other items, these features were often recorded, although some were broadcast live. Yet in *Picturegoer*'s discussion of the dangers of live appearances, this is constructed as a *new* problem associated specifically with television. Television's visual dimension rendered the stars' appearances much more exposed - as foregrounded by *Picturegoer*'s disdain for how the stars 'writhe nervously, held in enormous and frequently unflattering close-up'.[26] Television also begs a comparison with the film performance in ways that radio does not. John Ellis describes how the film performance permits 'moments of pure voyeurism... the sense of overlooking something... which passively allows itself to be seen' (1992, p.99) and this is related to the photo-effect of present-absence (p.100). (The person is there yet not there - while the image is in the present tense, the referent is in the past tense). Television coverage clearly shifted this temporality. Although a rhetoric of immediacy, direct address and an illusion of liveness continue to be central to television's address, this was perhaps particularly charged in the 1950s when the appearances were literally live, as this had particular implications for the construction and reception of the film star's performance.

In the case of the Crawford interview, the notion of 'current time' is made clear when Haigh introduces her by explaining: 'I think I should tell you it's her first appearance on television ever', and after greeting the presenter, Crawford immediately admits that she is nervous: 'Hi Peter, yes I'm scared'. This acts as a kind of commentary on her performance which draws attention to the here and now, the fact that it is occurring in 'real time', as we are watching the dialogue unfold. While she rejects notions of direct address and insertion into the conversational style of broadcasting which Marshall (1997) (and others) perceive as disrupting the aura and distance of the star, television's liveness still shapes her appearance in other ways. It may indeed be the case that the live nature of her performance undermines her 'auratic' status on some level. This is not related to the liveness of her appearance *per se* (in terms of mode of address, for example), but the way in which this makes her very nervous (Crawford had long since been terrified of making personal and public appearances) (Wayne, 1988, p.213). Her anxiety is clear from her body language. In contrast to the wide-eyed, defiant look adopted in her films, Crawford spends much of the interview with her gaze directed towards the floor, and only turns to acknowledge Haigh intermittently. She also wrings her hands, and when her husband Alfred Steele appears she clings to him in a nervous manner and is evidently grateful for his supportive presence. As he leaves the set Crawford continues her coy performance, holding Steele in her gaze until he is out of sight, and this all offers a stark contrast to the strong 'independent' woman image constructed by her films. It is hard to reconcile, for example, in terms of her recent film roles, the domineering Jenny Stewart in *Torch Song* or the ruthless Eva Phillips in *Queen Bee* with the more fragile, anxious 'real' Joan Crawford on *Picture Parade*.

Her nervousness is also apparent through her interaction with Peter Haigh. While he evidently feels that his meeting with Crawford is an honour, this is occasionally undercut by his interviewing technique. Haigh later explained that due to Crawford's fears it initially took four hours to persuade her to appear,[27] and his anticipation of her nervousness shapes the way he approaches the interview. When she pauses before responding to a question he sometimes prompts her, often before she has had a chance to answer. Particularly at the beginning of their talk, this gives the impression that Crawford requires encouragement and support, and that Haigh is himself nervous that she cannot carry off the interview. It is also a strategy for limiting her 'performance', as due to the time constraints of live television, he is eager to ensure the swift progression of the interview. Again, it is Crawford who is not accustomed to these conditions and while it is she that commands attention, it is Haigh who controls their exchange.

The interview again replays the ordinary/extraordinary paradox so central to cinematic stardom. In much the same way as a fan magazine, this is firstly constructed at a discursive level in that although the interview focuses on her career and talent, it is punctuated by references - particularly from Alfred Steele - to her domestic identity as a wife and mother, and Crawford's own advocation of a 'perfectly ordinary life'. It equally reflects the dominant construction of stars in the Classical Hollywood period which conventionally pivoted on a construction of their 'glamorous' lifestyle yet 'surprisingly ordinary domestic life of the star' (Geraghty, 2000: 184). (Furthermore, we can note here how television offers an ideologically conservative image of the star. Although her recent marriage to Steele had enabled Crawford to project a return to the 'stability' of family life, this is quite a shift from the previous reports on her behaviour in the scandal magazines focusing on drunken incidents and relationships). Yet on television this paradox is further layered at the level of performance. Joan Crawford is 'extraordinary' in so far as she has an excessive presence of which the presenter is clearly in awe (and she is talking about her 'exceptional' work as a film star), yet she can be seen to be 'ordinary' in that, like 'us', she can feel nervous and vulnerable, feelings which she cannot fully control. Viewers were invited to be 'in' on the 'joke' that appearing on television is still a new experience for many - whether famous film star or not. When considering how this paradox is played out in different media forms it is clear that in the 1950s, it could be shaped by the specificity of television and its aesthetic and cultural development at this time.

Although the interviews predominated where the focus on the star was concerned, stars also appeared in different forms of coverage which shaped their representation in different ways. The coverage of Joan Crawford was an extended version of what the *Picture Parade* memos described as 'Diary of Events', and with John Parsons referring to his star 'diary' each week, *Film Fanfare* also used the trope of the diary to narrativise the comings and goings of stars. Particularly on *Film Fanfare*, stars were often seen arriving at the airport (generally American actors and actresses), where they were situated in a more public space than that of the contained television studio. Nor were these appearances live, not least of all because *Film Fanfare* was only broadcast live for the first few weeks of its run.

'What is the Magic of this Fabulous Star'?: The Absent Monroe on *Film Fanfare*

In 1956 Marilyn Monroe arrived in Britain to work on the production of *The Prince and the Showgirl* (Laurence Olivier, 1957) in which she was starring with Sir Laurence Olivier. Her career had developed rapidly since the early 1950s and Monroe had graduated from playing small parts to starring roles. Films such as *Niagara* (Henry Hathaway,1952), *Gentlemen Prefer Blondes* (Howard Hawks,1953) and *How to Marry a Millionaire* (Jean Negulesco,1953) established her unique popularity, charm and appeal, and she was voted top female box office star by American distributors in 1953 (Dyer, 1986, p.27). By the time she appeared in *The Seven Year Itch* (Billy Wilder, 1955), her enormous earning power was clear. When Monroe arrived in Britain in July, 1956, she had just wed her third husband, playwright Arthur Miller, and the 'world was still buzzing with the excitement' of their marriage (Anderson, 1991, p.144). Monroe was to be in Britain for four months and prior to her arrival, there had been a huge build-up in the British press. The coverage constructed by *Film Fanfare* (and *Picture Parade*) was part of the phenomenal media attention Monroe received.

Shortly after her arrival John Parsons explained: 'This week I assembled a portrait on film of a girl who has taken London by storm. Here she is at the Savoy Hotel stepping into a blaze of light at the press reception'.[28] As we see the images, the camera is positioned as one of the crowd amongst the photographers and their flashing bulbs as Monroe appears in a tight, sleeveless black dress and takes a seat on the platform. As her blond hair is illuminated by the flashing lights, the viewer does not have a particularly privileged view and the camera retains a certain

Figure 15: Marilyn Monroe faces the media on Film Fanfare (1956, edition 23)

distance. The home audience is also positioned as one of the crowd, straining to get a glimpse of 'the dazzling star'. This works to emphasise the unattainability of Monroe, her 'exceptional' nature, and it maintains a certain distance between viewer and star. The camera then shifts its position and cuts to a medium close-up, and we see Parsons crouching down to the left of the frame as he asks the star some questions. (In the voice-over he recalls her statement that she was very excited about the new film). While this is more of a privileged view, her distance is still emphasised in so far as she does not acknowledge the camera, and we cannot hear her answers. We have to rely on Parsons' voice-over commentary as he relates *his* past experience of meeting Monroe. Arguably, this offers the incomplete image which Ellis suggests is so central to the cycle of desire and knowledge that characterises the consumption of the star (1992, p.93). Although Ellis claims that television does not promote the 'photo-effect' in that it 'pretends to actuality... [and] immediacy' (p.106), the concept is relevant here. In these early years it was not infrequent for filmed material to be silent, and when accompanied by a commentary (referring to the images in the past tense), this clearly differentiated it from live television. While these differences will be considered in detail in terms of the coverage of premieres (Chapter Six), the above example was marked as

Figure 16: Monroe waves to the crowd 'who weren't so lucky as the rest of us' (1956, edition 23)

existing in the past tense, and as having 'a very particular sense of present-absence' (Ellis, 1992, p.93), and this heightened Monroe's distance. In relating his past experience of meeting Monroe, Parsons necessarily mediates between viewer and star to a much greater extent than the live television interview.

The camera then cuts to a different viewpoint which positions 'us' behind Monroe in the press room as she leans out of the window to wave to the seething crowd below. As Parsons' commentary accompanies this image, he constructs television as having an ideal or privileged view and describes her 'wave for the crowds who weren't so lucky as the rest of us' - an 'us' which is intended to encompass the viewer as well. The material that follows is made up of library footage (which is in fact often used in television profiles of Monroe today). We first see the star singing to the GIs in Korea, then appearing at a Warner Bros studio party in Hollywood, and finally standing with Arthur Miller in the countryside following the announcement of their marriage. After the footage has run we return to the studio as Parsons enquires: 'But what is the magic that has put this star on the pinnacle of fame? We've been investigating the story behind the story and I've two men with me now who have special knowledge on the subject....' It is worth noting here how the construction of the stars in the 1950s cinema programmes also emphasises the shifts in what we might term cultural explanations of stardom and fame (Gamson, 1994). Drawing on myths largely developed in the earlier part of the last century, the programmes insist upon stardom as based on 'an undefinable quality of the self [as] natural, almost predestined' (Gamson, 1994: 32), and perpetually draw upon 'mystifying' terms such as 'magic', or 'an indefinable essence we call star quality'. Certainly, these discourses still circulate around the construction of fame today (see Holmes, 2004),[29] but as Gamson argues, they have increasingly been under attack by 'the challenge from the manufacture-of-fame narrative [which] has been greatly amplified... it has become a serious contender in explaining celebrity' (1994: 44). Gamson in part attributes this to the explosion of media outlets for image creation and industrial and cultural shifts in the 'post-glamour, television-dominated era' (Ibid). While this in itself may reflect back on prejudices toward televisual fame, it is significant to note here that in terms of film stardom on the cinema programme, television appears on the cusp of these shifts: resolutely fostering myths of fame developed in previous decades at a time when other media outlets (such as scandal magazines) were in some ways aiming to undermine their validity.

While in the example above *Film Fanfare* asks 'what is the magic that has put this star on the pinnacle of fame?' it also promises us 'the story behind the story'. Rather than the privileged visual intimacy promised by the camera's close-up, television is here constructed as an informative, investigative medium which is ideally suited to offering this access at a *discursive* level. The two men to whom Parsons refers were Bob Stannage, Director of Publicity for Warner Bros in Britain, and Earl Wilson, a famous American columnist, described by Parsons as 'a personal friend of Marilyn Monroe'. Bob Stannage was asked to tell the viewer 'exactly what happened from the moment the excitement started at London airport'. His response was constructed in narrative terms with the events related stage by stage, building up to a climax. This was presented from the film industry's point of view with Stannage explaining, for example, Warner's problems in finding a large enough room to accommodate the media. 'The story behind the story' is thus designed to give the viewer not simply 'inside' information on Monroe, but also on the orchestration of the event. This was similar to coverage of Monroe's visit in other media. The *Daily Express*, for example, had considered that the nation should know how London airport was proposing to cope with the arrival of the star and it interviewed officials of the airport and of the Ministry of Civil Aviation (Anderson, 1991, p.145). There seems to be almost as much emphasis on how London will accommodate the arrival of Monroe as on the star herself which indicates not only her enormous fame (how can we cope with such a phenomenon?), but - as with Crawford - the fact she is actually *in* Britain, rather than (the more geographically distant) Hollywood.

Just as aspects of the coverage work to construct Monroe as a distant and unobtainable figure, this is further played out in the interview in which she is quite literally absent. Unlike *Picture Parade*'s footage of Joan Crawford, the coverage was not followed by a personal appearance and it is rather a report *on* Monroe. Although Monroe's contract with Twentieth Century Fox permitted her to take part in a certain number of television shows (Anderson, 1991, p.136), she was one of the stars who did not appear on *Picture Parade* or *Film Fanfare*, something of which regular viewers would have been aware. In 1955 she was a guest on the American variety programme *The Jack Benny Show* (Mann, 1992, p.52), but it is doubtful that this was seen in Britain at this time. Indeed, the same *Picturegoer* article which ran the headline 'TV Wins Over those Frightened Stars' completed its statement with the qualifier 'but it can't get stars like Monroe'.[30] This referred to her refusal to appear in American television drama, but it similarly emphasised her unattainability where television was concerned. Paradoxically, while television is actively intervening in the promotion and construction of her image, her unavailability is also encoded in its textual form.

As with Crawford's appearance, this representation of Monroe is shaped by, and contributes to, the construction of her star image at this time. Richard Dyer's seminal work on 'reading' Monroe historicises her status as sexual icon, and although concentrating on this status, he suggests that other aspects of her persona - acting ability, intelligence - should nevertheless be recognised and recovered 'against the grain of her image' (1986, p.20). Dyer's own work of course argues that stars' images are polysemic and embody contradictory meanings, and while the emphasis on Monroe's sexuality may have been dominant, other more contested aspects of her star persona are not necessarily reading 'against the grain' of her star image, they are an active part of it. Monroe struggled throughout her career to be taken seriously as both a talented actress and cultured young woman (Anderson, 1991) and it is significant then, that with respect to *Film Fanfare*, Earl Wilson spends considerable time emphasising her intelligence and 'love of good books', insisting that she should not be judged as a 'dumb blonde'. Monroe had recently studied method acting during her year at the Actors Studio and for many critics, her role as Cherie in the comedy *Bus Stop* (Joshua Logan, 1956) had at last proved her to be a genuinely talented actress (Anderson, 1991, p.136). Richard Dyer refers to *Bus Stop* and *The Prince and the Showgirl* as Monroe's 'prestige roles' (1986, p.21) and her co-star in *The Prince and the Showgirl* was Laurence Olivier, obviously an actor of very high repute. Of further significance was her recent marriage to Arthur Miller when, as Janice Anderson explains, Monroe '"proved" her right to be taken seriously by marrying an admired member of the American intellectual establishment' (1991, p.136). The placing of 'proved' in inverted commas, however, suggests that many were not convinced - that it was a temporary, unstable, or non-existent 'victory'. Indeed, *Picturegoer* noted the contrasting backgrounds of Olivier and Monroe suggesting that in view of Olivier's reputation, the young actress had 'an awful lot to live up to'.[31] Olivier was also directing the film and he was apparently building a 'frustrating fence' around Monroe because he wanted to control her contact with the media. At the press conferences - including the one shown on *Film Fanfare* - questions had to be put to Olivier and only then was Monroe allowed to answer. *Picturegoer* suggested that 'Maybe he's scared that Marilyn, with her hitherto much-publicised reputation for frankness, might lower the dignity of the film he's making' and the magazine continued that due to Olivier's intervention her 'treatment at the hotel press receptions... was the sort you might give to the President... not a popular film star...'.[32] While this comparison with the President evokes associations of importance and prestige, *Picturegoer* is using it to suggest that Monroe was deliberately shielded from direct contact with the media. While this is of course only speculation, it is confirmed by the *Film Fanfare* material, and it is possible to suggest that tensions surrounding Monroe's star persona at this time may have actively shaped the aesthetics of the television coverage and the space constructed for the viewer. Perhaps Olivier's concerns over Monroe's 'hitherto much-publicised reputation for frankness' had as much to do with the visual and discursive distance of the television coverage as did her status as an 'auratic' being, and it gives a new inflection to her distance and unattainability.

Whatever the factors behind it, if Monroe's exceptional 'star' qualities, her 'aura', are constructed here through her absence, this clearly differs from the example of Joan Crawford. Crawford's aura is partly constructed by the fact that she is actually present in the studio, and that she is present *at that time*, as well as her uneasy relation with the communicative style and discourse of broadcast television. In contrast, the coverage of Monroe is organised not around presence but absence. In the filmed coverage, we are offered glimpses or 'fragments' of Monroe (the body but not the voice) and the image is incomplete. The interview takes this further. If the star replicates the dynamic of the photo-effect, representing an 'absence that is present' (Ellis, 1992, p.93), this is here played out on a different level to tantalising effect. Although physically absent from the programme, she is simultaneously constructed as a 'presence' through the offer of 'inside' knowledge, a discursive access mobilised through the discourse of 'intimacy and enigma' (Marshall, 1997, p.90). This is evident in the rhetoric of the feature, which asks 'What is the secret of this fabulous star?', then claims to answer this with 'the story behind the story' (which inevitably offers very little information at all).

The examples of Crawford and Monroe indicate how the construction, connotations and implications of the star's television appearance were dependent on the context of the coverage, as well as their particular cultural image. The question of whether television depletes the 'aura', glamour and mystique of the star is clearly a complex issue and - given that the answer is ultimately to be found at the level of *interpretation* - it represents a line of enquiry that remains more ambiguous than existing work suggests. There was certainly an attempt, particularly within the self-conscious discourse of television's developing modes of address, to draw the film star into the communicative context of informality and familiarity, but the extent to which this shift is accomplished is another matter entirely. Crawford and Monroe were of course both Hollywood superstars, and the very appeal of using their coverage here may indicate the extent to which the treatment of American stars could differ from that of British. The most prestigious, glamorous and exciting coverage was arguably reserved for Hollywood stars, or at least those with a high public profile, and the very concept of their geographical presence in Britain charges the coverage with the status of an event. The British stars were perhaps indeed treated more informally, although the question of how the 'aura' of the British star is constructed differently in the first place is clearly a wider issue (Babington, 2001). That is not to underestimate, however, the degree to which British female stars were also used by television as a resource of glamour and admiration. The next chapter explores the ways in which television used the possibilities offered by its own technological innovation to forge new relationships between the glamorous world of cinema and its domestic consumption.

Notes

1. 'TV Wins Over those Frightened Stars', *Picturegoer*, 7 September, 1957, p.13.

2. *Daily Sketch,* 12 June, 1956, BBC Press cuttings, 1955-58.

3. *Picture Parade* script no.44a.

4. 'You see - TV's only got six stars...', *Picturegoer,* 18 June, 1955, p.26.

5. For example, in 1953 an advert for Capstan cigarettes featured Jack Hawkins and referenced his current picture *The Cruel Sea* (Charles Frend, 1953). Hawkins insisted that 'after I've smoked this Capstan, I'll be ready to sink fifty U-boats!' (13 June, 1953, p.96).

6. *Kine Weekly,* 1 November, 1956, p.28.

7. BBC Viewer Research Report on *Picture Parade*. 12 August, 1957.

8. *Film Fanfare* no.29. Late 1956.

9. 'Too much praise is killing our shop windows' in *Kine Weekly,* 2 August, 1956, p.6.

10. BBC Viewer Research Report on *Picture Parade*, 23 June, 1958.

11. *Picturegoer*, 26 October, 1957, p.15.

12. 'Film Talks for *Women's Hour*'. 8 October, 1951. R51/173/5.

13. *Radio Times*, 8 February, 1957, p.5.

14. *Film Fanfare* no.9.

15. For example, if we were to apply the questions we ask of stars to the broader context of human identity, they explore 'useful' issues surrounding selfhood: is there a distinction between our 'private' and 'public' selves? Do we have any 'unique' essential, 'inner' self, or are we simply a site of self-performance and public presentation?

16. 'Has Hollywood Had It?: Part 2', *Picturegoer*, 26 March, 1955, p.18.

17. *Picturegoer*, 7 September, 1957, p.1

18. All analysis of the Crawford interview is based on *Picture Parade* (18a), transmitted 7th August, 1956.

19. *Picture Parade* 17a, transmitted 30 July, 1956.

20. *Picturegoer*, 24 August, 1957, p.8.

21. *Picturegoer*, 27 July, 1956, p.11.

22. *Picturegoer*, 26 May, 1956, p.8.

23. *Picturegoer*, 31 May, 1958, p.5.

24. *Picturegoer*, 27 July, 1956, p.11.

25. *Punch*, 28 August, 1956. BBC press cuttings box PP62.

26. *Picturegoer*, 26 March, 1955, p.18.

27. 'Picture Parade and *Movie-Go-Round*', *Radio Times,* 8 February, 1957, p.8.

28. *Film Fanfare*, no.23. July, 1956.

29. I refer here, for example, to the discourses of stardom constructed in pop Reality TV shows such *Pop Idol* (2001-2, 2002-3, UK), which combine these older myths of fame ('star quality', the 'X-factor'), with an open emphasis on manufacture and image construction (Holmes, 2004).

30. *Picturegoer*, 7 September, 1957, p.13.

31. 'Don't Fence Her in, Larry', *Picturegoer*, 11 August, 1956, p.5.

32. As above.

Chapter Six

Glamour, Showbusiness and Movies-in-the-Making: The Film Premiere and 'Behind-the-Scenes'

The coverage of the stars pervaded every aspect of the cinema programmes at this stage, not least of all in the reporting of film premieres, or items going 'behind-the-scenes' of filmmaking. Again, in terms of visual media, these are areas which were largely developed by television, and certainly continue to represent staunch elements in its coverage of film culture today. They are, then, key strategies in terms of how television has domesticated the cinema, and played a role in constructing its cultural meanings and circulation. In terms of the film premiere, although a consideration of the star is part of this section, a primary concern here is the *technology* used to transmit such events. The BBC made use of a live outside broadcast to televise a premiere, and this provides a further example of how film culture was shaped by the new technological possibilities of television. The chapter then considers television's construction of a 'behind-the-scenes' perspective on film production: the 'image' that it offered of filmmaking at this time and in particular, the industrial strategies of both British and Hollywood film companies to become involved in this process.

Livening up the Film Premiere

Despite their centrality in the media representation of the star, film premieres have not been the subject of much critical attention. This is perhaps because they are often seen as simply a promotional strategy and unworthy of further investigation. They were, and still are, publicity and 'gossip' for the media, a context which, from the late 1950s, also included television. As with many aspects of the cinema programme, television did not invent the coverage of premieres, and it was previously the subject of fan magazines, the press and perhaps most importantly, the newsreel. A crucial point of interest, then, is how television re-shaped or contributed to this material in its own particular way - and it is notable that aspects of this intervention were relatively short-lived. This was largely because of its symbiotic relationship with *live* broadcasting, which declined as television developed.

From the late 1930s to the 1950s, there emerged an enormous amount of critical commentary attempting to explore and define the social and technological uses of television. Critics compared television to other media - radio, film or theatre - often proclaiming a unique combination of their technologies, aesthetics and experience. Television's liveness, perceived as its crucial defining feature, was often compared to theatre, while its moving image was described as 'radio with pictures', a discourse which had an equally clear heritage in cinema. As these comparisons suggest, and as discussed in relation to *Current Release*, what was most clearly seen to differentiate film and television was live transmission, and this was used to map the differences between their aesthetics, address and audience experience. Typical of many perspectives at this time, Michael Clarke discussed this issue in the annual publication *The Cinema* (1952), arguing that television's liveness could generate 'the excitement of theatre and touch on regions where cinema cannot venture' (1952, p.182). Clarke's use of the metaphor of travel here, very common in the discussion of early television, implies not merely a difference between the media, but a difference based upon a spatial and temporal distance (Jacobs, 2000). Cinema and television take us on different 'journeys' in different ways and according to Clarke, engage the viewer in very different experiences. Much of the early commentary tended to theorise these differences on the broad level of technology and aesthetics, and ignored the reality of the developing institutional relations between cinema and television. However, this section considers not simply how the cinema *did* 'venture' into television - as the cinema programme clearly suggests more generally - but how television enabled a 'venture' 'to' the cinema in ways which make the metaphor of travel rather apposite.

Part of the aim of this book is to establish that it was not so much the feature film, but the cinema programme upon which domestication of film culture was forged. Within the context of television's increasingly domestication of film culture, a live outside broadcast of a film premiere is particularly revealing - literally bringing the cinema 'into' the home. It rehearses an intersection of public and private spheres, the merging of which, as Jacobs describes, was seen to be central to 'the novelty of television's modernity' (Jacobs, 2000, p.26). This then takes on a particular resonance in the context of the changing cultural significance of filmgoing, and of the cinema more generally in the 1950s. When considered in relation to the discourses on live television which specifically promoted the medium in terms of mobility and travel (Jacobs, 2000, p.25), the live coverage of the premiere perhaps enabled the term 'going to the cinema' to take on a new spatial and temporal meaning at this time.

To foreground liveness in the discussion of 1950s television is to risk accusations of technological essentialism whereby the 'early limitations [of television] define the "essence" of the medium, by which we continue to be bound' (Barr, 1996, p.53). William Boddy, for example, questions the early discussion surrounding television, and the discourses which envisaged its technological, aesthetic and cultural identity on live terms:

> The theoretical speculation which emerged was frankly amateur and produced a number of dubious and poorly-argued essentialist claims for the medium... Little of this quasi-theorising has been subjected to subsequent analysis, and this blindspot of television historiography... has contributed to the myth of the medium's 'Golden Age'

> *(Boddy, 1990, quoted in Barr, 1996, p.51).*

First and foremost this is a debate about the way in which the history of television is to be written. While it remains important not to 'fetishise the live element in television' and privilege this aesthetic as defining its formative history (Barr, 1996, p.53), it is equally problematic to dismiss the early discussion of television as having 'nothing much to teach us about the medium at all' (Ibid). Critics have increasingly found it useful to give serious consideration to the *discursive* construction of television in the 1940s and 1950s - the ways in which it was discussed, theorised and debated (Spigel, 1992, 1997, Oswell, 1999), and given the paucity of audio-visual footage from television at this time, such evidence is crucial in understanding the ways in which the future and possibilities of the medium were conceived. They are less 'naive' or 'inaccurate' prescriptions of the key qualities of television, or uninformed forecasts of how it would develop, than they are vital in understanding prevailing institutional and cultural perceptions of television in its formative years. (Barr 1996). This is central in attempting to reconstruct the live coverage of the premiere, which does not exist in audio-visual form. To draw again on Jason Jacobs' description of the 'ghost text' as a way of understanding early programming which primarily exists in written discourse and evidence (2000, p.14), the coverage of the premieres might be understood as 'ghost events' (in that they partly subordinate the television text to the relaying of an 'event').

There is a sense in which working from written material is restricting and inadequate, with the entire concept of the live premiere being based on the immediacy of the visual, and its consequent experience. At the same time, at a methodological level, it is questionable whether an audio-visual record would be entirely appropriate. The ephemeral nature of these images, their liveness, is in itself evidence of how they were intended to be transmitted and experienced: it is the integral marker of their historical significance. Part of my interest here is to compare the live and filmed premieres. Where the latter are concerned, it is possible to work from available footage, but the analysis of the live premiere remains overwhelmingly discursive. From the researcher's perspective, meaning, then, must partly be sought in the discourses and conceptions surrounding live television. This can then be linked to discussions of the live premiere in the press and trade reports, or in the BBC scripts and memos. In its own way, this offers a fleeting glimpse of its existence, and the possibilities of the premiere in terms of aesthetics, form and experience.

In terms of the more general footage of film premieres, it is clear from the correspondence with the film industry that the BBC had hoped to cover such events earlier in the decade when negotiating over *Current Release*, but no regular material emerged at that time. In 1952 and 1954, premieres were part of the programmes called *Italian Film Festival* which covered two festivals of Italian films held in London. Evidence exists in relation to the second event in 1954, and it indicates that it was a very glamorous affair. The festival was held at The Strand's Tivoli Cinema in London between 25 and 31 October, and the BBC offered two programmes covering the event. On Monday 25 October 1954 (7:30pm-8:00pm) they screened an outside broadcast of the foyer scenes in the cinema which featured not only the stars, but also the arrival of the Queen. The memos emphasise the importance of timing in this respect ('her Majesty has in the past been most co-operative in her understanding of the timing of BBC programmes'),[1] so it is possible that this broadcast was live, prefiguring the later example in *Picture Parade*. Two days later (10pm-10:45pm) the BBC broadcast a general programme on the festival including interviews with its stars and directors, fashion parades and film extracts.[2] The first televised premiere organised around a specific film was that of *Kings Rhapsody* (Herbert Wilcox, 1955), a historical costume drama starring Errol Flynn and Anna Neagle.[3] It is unclear on which channel it was screened, but it appears to have been transmitted on film. The coverage of these events was thus still relatively uncharted territory by 1956 and as a result, there was much excitement when as the focal point of *Picture Parade*'s second edition, the BBC intended to cover a film premiere.

Carousel to *The King and I*: Experimenting with the Film Premiere on *Picture Parade*

This was to centre on the British opening of the American musical *Carousel* (Henry King, 1956). The *Carousel* premiere was not to be transmitted live, but re-broadcast one hour after the event had occurred. But this occasion was important for a variety of reasons. Firstly, it was widely perceived as the first true link-up between film and television in the coverage of a film premiere and as such, it attracted an enormous amount of publicity. Secondly, *Carousel* was produced by the Hollywood major Twentieth Century Fox, and was part of the company's wider strategy of experimenting with the new technological possibilities of television, as the discussion of the live premiere will make clear. Perhaps most importantly, while the *Carousel* event was transmitted on film, it drew on aspects of liveness at a discursive level. With the coverage transmitted only one hour after the occasion had taken place, the *Carousel* event straddled the boundary between the live and the filmed premiere.

With their pre-sold appeal and lavish aesthetics, 1955-6 saw several adaptations of such giant musicals, including *Oklahoma!* (Fred Zinemann, 1955) *The King and I* (Walter Lang, 1956), and *Guys and Dolls* (Joseph L. Mankiewicz, 1956). In 1954 Twentieth Century Fox announced a shift in their production strategy from stars to subject matter, with films that would emphasise the advantages of CinemaScope (Maltby, 1995, p.156), and while they still drew on star appeal, such musicals fulfilled this role. From the trade's point of view they were considered to have considerable earning potential and *Carousel* was further promoted as an event in that it was the first film produced in the latest widescreen format, CinemaScope

"55", an aesthetic innovation complimented in the reviews.[4] While the American release of *Carousel* actually lost money for Twentieth Century Fox, *Kine Weekly* claimed that it was generally successful at the British box office.[5] The critics considered the film to have a wide appeal and as one review noted, 'Here's a great show for practically anybody',[6] a sentiment reinforced by the trade press. Given its status as a family film and its poor returns in the US, Fox perhaps intended to promote the film particularly heavily in Britain, a campaign in which television was required to play a key role.

Discussion not only surrounded the new CinemaScope "55" process, but also what the trade described as 'the sparkling new singing star', Shirley Jones. Although Shirley Jones and Gordon MacRae had previously been paired in *Oklahoma!*, this was yet to have its British release, and *Carousel* was Shirley Jones' British debut. Much discussion surrounded her performance and new star identity, and Jones received an enormous amount of publicity in the press and popular magazines. The *Daily Sketch* was particularly appreciative and in a feature article on the actress, generously hailed her as the new Grace Kelly.[7] Certainly perceived by the trade press as one of the biggest films of the year, these various factors contributed to the excitement surrounding *Carousel* at the time of its British premiere. Twentieth Century Fox organised the event in conjunction with *Picture Parade* - the BBC were to film the events at approximately 9:00pm, and they would then be re-broadcast in *Picture Parade* one hour later. It was the *Carousel* coverage which opened the programme, beginning with an address from Peter Haigh:

> Good evening viewers and welcome to Picture Parade number two. It has been an exciting night in London's West End. The European premiere of the Drury Lane musical... Carousel has taken place and Shirley Jones, the American singing star, has flown from Hollywood. Picture Parade cameramen went along to cover the event. Many personalities of the entertainment world were there, including someone well known to viewers - Dennis Lotis.[8]

Lotis was a popular singer (who appeared on television and had also acted in minor roles in film), and Fox had specifically boasted how the 'guests of honour will include all the top singing names in television and theatre', including stars such as Alma Cogan and Vera Lynn.[9] Lotis was also to present the feature on *Picture Parade* and while this decision was fostered by his link with the event, he was also standing in for Derek Bond. Lotis then entered the studio and 'ad libbed' with Haigh, before walking toward the 'office' and sitting at the desk, with the row of monitors behind him. The viewer was then shown silent sequences from the premiere scenes including the excitement of the guests arriving and the interviews in the foyer, overlaid by Lotis' live commentary from the studio (no details are given in the script but it also included 'music and effects').[10] The camera then cut back to Lotis at his desk as he explained 'Before we meet *Carousel*'s lovely young singing star, Shirley Jones, let's see her singing a number from the film, 'What's the use of Wonderin''. Jones then arrived on the set straight from the premiere and was interviewed by Lotis. The interview was interspersed with two further clips from *Carousel* including the musical showpiece of the film, 'June is Bustin' Out all Over'.

Fox were eager to issue a press release to publicise the event, and they foregrounded the media synergy which made the coverage possible:

> The gala premiere of *Carousel* .... will be the first true, full-scale television premiere ever mounted by a film company. It is the first occasion on which a company has planned its premiere in conjunction with BBC Television to fit their pro-gramme...20th Century-Fox [sic] planned the date, time and presentation of the pre-miere in conjunction with the BBC and have given full facilities for television cover-age.[11]

While seeking to highlight their pioneering spirit, this release references the understanding that, in other areas, the media are still on hostile terms. It functioned as an address to the film industry encouraging them to see the benefits of further co-operation, while simultaneously emphasising the contemporary and 'progressive' nature of the company's approach. The implications of this move were later noted by the *Evening News* (in connection with the live premiere of *The King and I*) when it ran the headline:

> Film Men Get in Step With TV: Bow down before the power of television.... a film premiere has been changed to fit a television programme. To realise the full import of this you must remember that film men... do not even change the dates of pre-mieres to suit each other.[12]

This, however, seems to miss the point - that it was an event intended to work for the mutual benefit of both media. Far from 'bowing down', it was clearly a calculated move by Twentieth Century Fox to experiment with the new possibilities of television. Coming so early in the development of *Picture Parade* (and specifically promoted at the end of the first edition), it was similarly a calculated bid by the BBC to differentiate their programme from *Film Fanfare*.

Twentieth Century Fox claimed to be 'pioneers in film publicity [and]... in new forms of presentation'[13] and their experiment with *Carousel* followed the pattern of similar coverage in America. In 1952 the first national telecast of an American film premiere centred on Fox's *Stars and Stripes Forever* (Henry Koster, 1952).[14] The *Motion Picture Herald* reported that it would 'undoubtedly suggest tie-ups for other films. This is one type of co-operation between television and [cinema]... that can only benefit both media, and the public as well'.[15] Four years later in 1956, such tie-ups were likely more routine in America, and it is an example of the extent to which Hollywood's methods of publicity are adopted in the British context (see Chibnall and Burton, 1999). Although the British film industry were very much involved with the coverage of premieres, it is logical that it would be an American company, and Twentieth Century Fox in particular, who would introduce such possibilities to Britain.

Yet the *Carousel* event was not broadcast live, and the coverage was explicitly marked as existing in the past tense. Haigh begins by explaining that 'It *has been* an exciting night in London's West End... [my italics]', while Lotis is seen simultaneously in both the coverage and the studio - a physical impossibility which marks the different temporal status of the footage. Again important here is the use of silent material with a commentary which, as in the Monroe coverage on *Film Fanfare*, clearly marks the material as 'then', rather than 'now', depending as it does on a recollection of the event. With its direct address in the live studio, this constructs a distance between the spatial and temporal origins of the commentary and the events depicted on film. There was also a tension with the extent to which the event capitalised on a rhetoric of liveness. This was not so much in its textual address, but in the discursive hype surrounding the feature (which again points to the crucial importance of written material here). Fox's description of the occasion as organised to 'fit' *Picture Parade* explicitly emphasised their close temporal existence, while Shirley Jones' arrival straight from the premiere is key in this respect: the fact that it had just happened was clearly promoted as part of the coverage's appeal. This was perhaps best summed up by the *Daily Film Renter* when it noted how it 'cashed in on the red-hot topical value' and 'excitement' of the scene.[16] The trade paper suggested, however, that it might have been preferable if the premiere was literally broadcast live, something the BBC were soon to experiment with when covering the British premiere of *The King and I*.

At a methodological level, it is here that the discourses surrounding live television, and the outside broadcast in particular, are of considerable importance. In the pre-war and immediate post-war years, the live nature of television was described through particular tropes and conceptions. As Jason Jacobs explains, 'Early television... was... promoted in terms of mobility - the transport of images *to* the home, to the invitation to journey *from* the livingroom ... to distant events and locations' (2000, p.25). The latter conception described above was often associated with television's identity as a relay device - 'passively' transmitting a pre-existing event such as sport, for example - and the live outside broadcast was central here. In fact, in the 1930s and 1940s it was apparently the live outside broadcast that was perceived to define the medium's technological and cultural function, at least by those involved in discussing its development. A report to the Television Committee in 1943 emphasised that it was 'the televising of actual events... the ability to give the viewer a front-row seat at almost every kind of exciting or memorable spectacle' that would be the medium's 'greatest service' (Briggs, 1979, p.188). In 1950 this was still confirmed by John Swift's *Adventure in Vision* when he explained that 'it is an accepted fact that the outside broadcast "sells" receivers in the first place' (p.126). Swift's book was one of the first texts to differentiate between the arts and techniques of television and those of film and it was the outside broadcast that was used to map these distinctions. The Assistant Head of Outside Broadcasts made a similar comparison, and paid particular attention to the role of the viewer:

> The job of the Television Outside Broadcasting Department... is to bring to viewer's screens as many as possible of the interesting public and sporting events at the very moment they are taking place. It is this spontaneity which makes a television outside broadcast more exciting than either a studio item or film. The sense of immediacy and intense realism of an outside broadcast cannot fail to grip viewers, who although perhaps many hundreds of miles from the event itself, may find themselves clutching the side of their armchairs with almost as much excitement as it they were actually present [17].

In linking the 'sense of immediacy' to an 'intense realism', Dimmock constructs the outside broadcast as a specifically authentic kind of liveness, a visualised form of physical presence. The live outside broadcast, then, was clearly seen to epitomise the second conception described by Jacobs above in which the invitation is 'to journey *from* the livingroom [original emphasis]' (p.25) to a space or location 'out there' in the public sphere. Lynn Spigel describes this as 'a sense of "being there", a kind of *hyper-realism* [original emphasis]' (1992, p.14). While the trope of collapsing the distance between viewer and object is crucial here, aspects of its novelty seemed to be in experiencing this visual mobility while *remaining* in the domestic comfort of the home. A letter in the *Radio Times* in 1952 encapsulated this when the reader enthused 'Thankyou for the lovely trip to [the racing at] Twickenham on Sunday... seated by the fire in my own home... it was just perfect'.[18] The outside broadcast did not necessarily abolish an awareness of spatial separation, and this was perhaps one of its pleasures. Nevertheless, when relating this to the premiere, if the live outside broadcast enabled the audience to 'share the excitement and the sense of being.... at great occasions with those actually present on the spot' (Miller-Jones, 1948, p.46), then the BBC perhaps offered the viewer the chance to 'attend' this event as it happened.

With respect to 1956 when the BBC pioneered this coverage, the importance of live material, and of the live outside broadcast in particular, had perhaps shifted somewhat. As noted previously, as the 1950s progressed, television made increasing use of filmed material on a number of different levels, weakening the validity of the claims that championed liveness as its essential aesthetic and technological basis (Barr, 1996, p.50). In particular, ITV ushered in a noticeable increase in filmed material. This was not necessarily, however, a decisive shift. The extent to which liveness could still be perceived as the definitive aesthetic of television in 1956 is suggested by the complaints in the press regarding the increasing screening of feature films, complaints, for example, which enquired 'Is TV a live medium or is it a mirror for the products of the film studios?'[19] Although the figure may not be entirely accurate, *Kine Weekly* noted in 1956 that approximately 80% of the BBC's material was live,[20] and the outside broadcast continued to be important within this context. (For example, recalling her work on the *Tonight* programme in 1956, Grace Wynham-Goldie described the outside broadcast as 'the essence of television' (1977, p.210)). So while the coverage of the premieres did not emerge from the discursive context of the earlier perceptions of television, its live identity was by no means institutionally and culturally obsolete. We should also remember that many families were only just purchasing their sets at this time (see O'Sullivan, 1991) and although it is important to acknowledge the early existence of 'guest' viewing at a friend or relative's house, this necessarily shapes the extent to which television, and its forms and aesthetics, are experienced as a novelty.

The premiere is specifically conceived as an evening occasion, a glamorous night out. Lynn Spigel suggests that rather than simply being an agent of privatisation, television was used to 'mediate the cultural transition from public to private entertainment' by offering imaginary 'nights out' (1997, p.226). The television network NBC claimed that 'In our entertainment, we... start with television as a communications medium, not bringing shows into the living room... but taking people from their living rooms to other places - theaters [sic] ... movie houses, skating rinks, and so forth' (Spigel, 1997, p.226). While the premiere may have offered an imaginary night out, it is different from these other examples as it partly offered a new form of access to film culture - something not generally experienced before. Unlike the cinema or the theatre, the film premiere was not generally accessible to the public. Although there had been coverage of premieres in the press, fan magazines and in the newsreel for some time,[21] television made these events even more accessible to the audience. This is illustrated by a letter to *Picturegoer* in 1954 which complained that 'Too often attendance [at film premieres] is limited to trade and society personalities and picturegoers have to wait until the following day to see the show. Surely more publicity would result if serious cinemagoers were invited to the premiere?'[22] Written before television began its regular coverage of the premiere, the reader appears to reference those shown in the cinema newsreels. While clearly not statistically reliable in foregrounding public perceptions of this issue, the letter is interesting in its emphasis on the temporality of the event as an important part of the experience. Having to wait until 'the following day' is seen as less satisfying than witnessing the occasion as it happens - engendering the feeling of being part of the event.

By the time the BBC featured the live outside broadcast in August 1956, filmed material of premieres was becoming more routine, and the live coverage was perhaps again an attempt to differentiate *Picture Parade* from *Film Fanfare*. A live outside broadcast would have been an impossibility for the rival programme given that from April 1956, it was transmitted on film. It was also not surprising that it was the BBC that experimented with these possibilities given that, developing the outside broadcast first in radio and then television, the Corporation had greater experience in this area. In these early years of competition when ITV persistently achieved a greater popularity, the outside broadcast was still an area in which the Corporation continued to excel (Briggs, 1995, p.22). Like *Carousel*, *The King and I* was acknowledged to be an important 'event' film by the trade press and in Britain, it was Fox's most successful CinemaScope picture of 1956.[23] Now considered a classic, *The King and I* starred Deborah Kerr and Yul Brynner, and as a new star, Brynner attracted considerable coverage in the press and fan magazines. He had received critical acclaim for his performance in the stage version of *The King and I*, and *Variety* felt that his screen exposure would make him 'an international personality'.[24] It is significant that, like *Carousel*, *The King and I* was generally considered to have an extremely broad appeal. As *Variety* insisted, it would draw 'audiences of both sexes, of all ages, and of all tastes'.[25] Again, this perhaps made it particularly appropriate for extended coverage on *Picture Parade*. As a build-up to the live premiere, Deborah Kerr had already appeared in the programme to sing one of the (dubbed) musical numbers and in a subsequent edition she was seen in a filmed interview.[26] The film was also previewed in *Film Fanfare* and included in *Close-Up* as part of two special programmes on Twentieth Century Fox. It is thus interesting that a critic in the *Daily Herald* hailed the film as 'the real and only true answer to television. It has breath-taking colour, splendid settings, and a glorious sweep, which is going to leave thousands of home screens idle whenever it's in the vicinity'.[27] While this may have been so, it is possible that television played an important role in filling the cinemas in the first place. Furthermore, it was through the film programme that viewers may have experienced the most 'glamorous' and prestigious glimpses of *The King and I*, at its live London premiere.

Again issuing a press release in anticipation of the event, Fox proudly announced that 'Many stars and celebrities will attend the premiere... which has been timed to be on the air "live" in *Picture Parade*...'[28] Haigh introduced the feature in the studio as the images appeared on the monitors behind him: 'As you see, the foyer is crowded with many interesting personalities and waiting to introduce them to us is Derek Bond'.[29] Haigh's address here is based on the notion of co-presence, and clearly differs from the temporal and spatial relations between studio and event constructed by the example of *Carousel*. As the camera zooms in on the screen behind Haigh displaying the images of the scenes, the monitor functions as a 'window' into the public sphere - in this case of show business - enabling the viewer to see its simultaneous existence from the privacy of the home. Situated in the foyer of the Carlton Theatre with its blooming flower arrangements and walls adorned with elaborate paintings, Derek Bond then describes the scenes and interviews many of the celebrities present. Rather than being more of an observer of the scene the audience is addressed directly by the presenter and invited into the glamorous space - an address impossible when the footage is on silent film, overlaid with a commentary.

In the discussion of the live outside broadcast, the viewer is often described as being placed in the position of witness, as though they are actually present at the event. In relation to the outside broadcast of the premiere, there is certainly something very appealing about this conception - viewers making 'virtual' trips to the cinema when actual filmgoing was in decline. Yet my description above of a 'window' into the public sphere partly references the other conception of television's live experience described at the start: as an 'act of delivery' in which the images are transported into the home, rather than the viewer being transported to the event (Corner, 1991, p.12). The direct address of Bond also positions the viewer less as a witness, than as *audience* (I am here, and will tell you what is happening). Based on the limited evidence available, the address and rhetoric of the coverage seems to shift between these positions. The *Daily Film Renter* reported that an 'estimated 11,000,000 saw premiere scenes' and it published pictures of the event depicting the outside broadcast unit at work in the foyer.[30] They also included Derek Bond interviewing a sparkling Joan Collins, as well as chatting to Darryl Zanuck, former Production Chief at Twentieth Century Fox, and his wife.

The trade press also discussed the technological innovations which made this broadcast possible, and which had a considerable impact on its aesthetic construction. Although in the immediate post-war period the live outside broadcast was seen as the medium's definitive function, it was only from 1952 that mobile cameras could be moved around at short notice (Briggs, 1979, p.869). In 1954 there were further developments with a more sophisticated mobile camera which was initially used to cover sport and could follow action over a fairly wide area (Briggs, p.869). Known as the Roving Eye, it came into general use in 1955. Briggs describes how 'It at last made it possible to cover events at comparatively short notice and to transmit pictures while on the move... it was possible for a Roving Eye single-camera unit to *revolutionise what the viewer could actually see*' [my italics] (1979, p.870/p.997). The Roving Eye was used for the premiere of the *King and I* and Fox used its new possibilities to promote the feature, mentioning it in the press release. The camera was also used for other entertainment programmes such as the BBC's *Saturday Night Out*. Again drawing on the 'intense realism' associated with the outside broadcast the *Radio Times* exclaimed, 'Come with the man with the mike! See things happen! When they happen! Where they happen!'[31] The possibilities of the Roving Eye were seen in the first edition of *Picture Parade* when (in connection with his film, *Charley Moon* [Guy Hamilton, 1956]) Max Bygraves was interviewed at the stage door of the Hippodrome Theatre where he was currently performing. *Today's Cinema* acknowledged its visual mobility when praising the programme's 'less static methods' of presentation[32] in comparison, for example, to the studio interview, and *Kine Weekly* hailed the coverage as a productive form of media synergy when it noted that the 'free use of outside broadcasts suggests the wider scope of the kinema [sic]'.[33]

These comments are particularly interesting in the context of the critical commentary in which television's liveness was described as differentiating it from film, enabling it to generate 'the excitement of the theatre and touch on regions where cinema cannot venture' (Clarke, 1952, p.182). Despite using the word 'cinema', Clarke is actually referring to viewing a film, but the comments from the trade press nevertheless indicate ways in which television's liveness could intermesh with film culture, again cast in terms of a spatial interaction ('less static methods of presentation', 'the wider scope of the kinema'). The *Daily Film Renter*, however, appeared to disagree with the other reports, and its response is interesting in considering the specificity of the televised coverage in comparison, for example, with that of the cinema newsreel. The critic described *The King and I* event as nothing more than a 'dignified scrimmage' in which 'there was too much noise to be heard', making it difficult for the viewer to focus on the stars. Those interviewed seemed to be 'yelling back at [Bond]' and the critic ended on the note that 'The lesson seems to be that careful planning with the stars beforehand should be the order of the day when the TV units go out to a premiere'.[34] In contrast to the other papers, the *Daily Film Renter* responds negatively to what are essentially indicators of the live 'authenticity' of the event, particularly in terms of its impact on the representation of the star. 'Careful planning beforehand' is clearly opposed to spontaneity and liveness. Rather than the star being carefully positioned, framed and lit (as we would expect in a film and to a lesser extent, in the television interview), their appearance is subordinated to the technological possibilities of the medium: the Roving Eye, and the excitement and atmosphere of the occasion. This also emphasises how the live outside broadcast was not simply structured by a neutral, 'transparent' relay function (Barr, 1996, p.51), as the technology actively shapes the aesthetic and form of the event.

It would seem that in the early stages of *Picture Parade*, certain items were intended to showcase the technological possibilities of television, and the pioneering role of the BBC in this respect. The first edition included not only the live outside broadcast from the Hippodrome, but an interview with the French director Christian Jacque, via the Eurovision link with Paris (and the Eurovision was also used in the 1954 programme *Italian Film Festival*).[35] Although the link returned in the celebratory 100th edition, it was not used regularly on *Picture Parade* and as with the live outside broadcast interview, its inclusion in the first edition suggests a novelty factor - effectively demonstrating what television could do. That is not to suggest that the live outside broadcast was still a novelty in 1956, but what is new is that this technology, whether the live outside broadcast or the Eurovision, offered new ways of experiencing film culture. These possibilities were still novel enough to be an attraction in their own right and the live outside film premiere reflected this logic. Part of its novelty perhaps also lay in the fact that it brought the very public face of the cinema (premieres are essentially about orchestrating the visibility of a film's opening under the public media gaze) into the home at a time when the domestication of film culture was still in its earlier stages. As such, it represents a particularly interesting example of how television mediated its coverage of the cinema which both epitomises and reflects upon these early stages in medium's development.

'Lots of Excitement, Glitter, Glamour': Plotting the Film Premiere

Not every company was willing to change the time and date of their premiere in order to secure live television coverage, and from the BBC's point of view, the live outside broadcast involved a great deal of planning and technological resources. The remaining premieres were transmitted on film, a technological difference which re-shaped their construction and presentation. For example, *Picture Parade* used film to record the premiere of *The Eddy Duchin Story* (George Sidney, 1956) and Peter Haigh explained how 'Last Saturday Columbia gave a midnight matinee of their new film all about show business... Nearly every star in London attended'.[36] Again shot in CinemaScope, the film told the true story of a young boy whose talent for the piano catapulted him to stardom, before his tragic death, due to illness. *Picture Parade* used this theme to narrativise the construction of the event, for as Haigh explained: 'On your behalf, [we]... sent along one of the many hundreds of young people in Britain who, with ambition and courage, are trying to make it to the top'. Described as a 'charming girl', this was Dorinda Stevens who was currently playing a minor part in the production of Charlie Chaplin's *A King in New York* (1957). The premiere footage featured Haigh escorting the glamorous starlet to the event. It was presented through carefully edited shots which show the couple catching a taxi, travelling to the cinema, arriving at the premiere and finding their seats. Back in the studio Haigh explained how 'I told [Dorinda]... that I hoped one day we'd be showing one of her films on *Picture Parade* ... In show business, the breaks really do come when least expected'. While still exploiting the glamour attached to the premiere, its representation on film was more constructed. In contrast to the pervasive mobility of the Roving Eye, this may have offered a more polished aesthetic, foregoing the authenticity of 'being there' admidst the action. According to critics of the time, viewing an event on film substantially altered the experience of the audience. Directly referencing a more polished aesthetic, critic and documentary filmmaker Michael Clarke described filmed material as 'refined, approved, rubber stamped' (1952, p.182). In the *Penguin Film Review* Andrew Miller-Jones argued that 'To see a record of something which has happened and to witness something which is palpably occurring before your very eyes are two entirely different matters. It is the difference between belief and make-believe' (1948, p.46). A recorded programme erected what Miller-Jones referred to as a 'barrier' between event and audience, it robbed the viewer of 'the thrill of knowing that the performance is taking place at that instant' (p.46).

When transmitted on film the events were plotted and narrativised for the viewer, and *Film Fanfare* used a similar structure to that of *Picture Parade* when covering the premiere of the adventure film *Safari* (Terence Young, 1956).[37] Jock MacGregor (who worked in the publicity side of the industry and claimed to attend approximately fifty premieres per year) presented footage of his trip to the premiere with the aspiring starlet Margaret Simons - soon to appear in the British film *The Silken Affair* (Roy Kellino, 1957). Leading into the coverage, MacGregor's voice-over accompanies a montage of shots from previous occasions and the item was essentially an introduction to the function of premieres, particularly with regard to aspiring starlets such as Margaret Simons. In an excited and breathless tone MacGregor explained that 'Film premieres mean lots of excitement, glitter, glamour. The fans love them and for many of the artists they are pretty nerve-wracking. They're on parade, and must look their best...' The footage shows MacGregor introducing Simons to various celebrities while his commentary explains: 'You've probably never heard of her but she may be a star of tomorrow. She wouldn't be the first star I know who has been helped by attending a premiere...' Again, this was a means of structuring the event in narrative form which gave the viewer a point of reference in a dizzying array of sparkling stars. Furthermore, Margaret Simons provides a point of identification for the viewer in so far as, like Margaret, the viewer is positioned so as to have the functions of the premiere explained to them. After the footage she describes her experience of the occasion and explains how she 'had a wonderful time. So many bright lights and cameras'. Incidentally, we can also see a gendered inflection here. On both *Picture Parade* and *Film Fanfare* the situation is structured by, and told from the perspective of, an older, 'wiser' and more worldly male introducing a young, 'naive', female ingénue to the daunting world of show business. Apparently, this even required Jock MacGregor to lecture the young Margaret Simons on the importance of female propriety in the world of show business. She is advised to attend premieres but 'not too many', lest she get a 'reputation as a girl who is always seen at premieres, but never on studio stages...' This could equally be interpreted as indicative of the extent to which British cinema is ill at ease with the excess of show business and showmanship, and this raises the question as to how the British film industry participated in this coverage. This perception of the British industry would seem to support Paul Swann's claim that:

> The American film industry was much more likely to stage special events where the celebrities could be seen and admired than the British industry. This was... something which the British film industry learned to mimic. The first Royal Film Performance took place in November 1946, and was really the first attempt to replicate the form and experience of the Hollywood premiere... Interestingly, it was very poorly orchestrated

(1987, p.73).

This perception of a British film industry as being less sophisticated and more uncomfortable with exploiting the promotional aspects of film culture is partly supported by the article 'Premieres - are they worth it?', published in *Picturegoer* only days before the cinema programme returned in 1956. The article explored a range of views on the value of the premiere including those of directors, publicity agents and stars, and the British and Hollywood industries are again perceived to have differing approaches to the premiere. Producer-director Roy Boulting expressed the view that: 'It's all Americanization... ballyhoo, trumpets and arc lights pay off over there. But British films are aimed at a different and perhaps less gullible public'.[38] A publicity agent presented this aesthetic in a more positive light suggesting that 'There's not enough of the Hollywood outlook here. You've got to bang... the big drum of publicity'.[39] Such discourses can be seen to be shaped by a particular perspective on British cinema on a number of different levels, and suffice it to say here that 'showmanship' has long been held in low regard by intellectual film culture in Britain. As Burton and Chibnall argue, typical British restraint and the desire for a realist 'quality' cinema were contrasted with the 'conventional "ballyhooliganism" of Hollywood' (1999, p.83). The television coverage of the film premieres, however, was at odds with a British cinema resisting the glamour and excess of the occasion. This is best demonstrated by a particular aspect of this coverage: the representation of the female star.

'Isn't She Lovely? Come on in for a Close-Up': Fashioning the Premiere

Following Charles Eckert's seminal article 'The Carole Lombard in Macy's window' (1978), film historians have frequently emphasised the relations between consumerism, cinema and the female viewer, ultimately producing what Barbara Klinger refers to as a 'small cottage industry' on the subject (1997, p.119).[40] The cinema screen has been conceived as analogous to a shop window in that it promoted not simply the stars and a particular consumer lifestyle, but a whole array of products from clothing to cosmetics. Such work has primarily focused on Classical Hollywood cinema and American female audiences. Jackie Stacey's *Stargazing: Hollywood Cinema and Female Spectatorship* continues the focus on Hollywood stars but investigates their appeal to British women in the 1940s and 1950s. Using women's memories of this period (obtained via an advert placed in a magazine), she suggests that Hollywood stars appealed to British women partly because they transgressed the conventions of a more conservative and 'restrictive British femininity' (1994, p.198). Stacey's *Stargazing* is now a widely referenced and impressive work, yet in its focus on Hollywood ideals, it has arguably perpetuated the impression of difference between British and Hollywood stars - in particular that British female actresses were not associated with such glamour, nor the consumer practices surrounding its construction. Yet it would seem that television could function as a smaller, although equally powerful, shop window where the film star was concerned.

It is true that the cinema programme aimed for a broad audience appeal but that, at certain points, the possibility of an address to a female viewer is evident. Whether in the figure of the modern working woman ('Susan') in *Current Release*, *Film Fanfare*'s initial intention to explore items on fashion and home decor, or the limited evidence indicating the sexual appeal of the presenters, it is possible to speculate about the existence of either a gendered address or audience. This section discusses what is perhaps the most visible aspect of this relationship in terms of the female star, fashion and the film premiere, specifically based on footage of *Film Fanfare*. Although there is not space to discuss the context here, this needs to be understood within the broader range of television programming which addressed itself to the female viewer, particularly concerning fashion, appearance and consumerism (see Leman, 1987, Thumim, 2002).

In the various premieres featured in the existing editions of *Film Fanfare*, the stars differ in the extent to which they address the camera directly. When the camera is positioned in the cinema foyer and the stars are entering through the doors, the stars sometimes address the viewer with a fleeting glance or smile, or (in many cases) not at all.[41] Particularly striking, however, is the way in which a female star will occasionally deliberately pose for the camera, and it is worth noting here that the structuring of this spectacle seems to be intrinsically linked to the aesthetic of the filmed coverage. If we recall the *Daily Film Renter*'s description of *The King and I* event as nothing more than a 'dignified scrimmage',[42] the differences become clear. A 'scrimmage' is not the place for the female star to parade and display her attire, nor for the camera to capture this spectacle. To take the example of the *Yield to the Night* (J. Lee Thompson, 1956) premiere shown on *Film Fanfare*. As the camera films the succession of guests entering the foyer, it pauses on the celebrity Sabrina, who made a name for herself by appearing on television as the 'beautiful blonde' who didn't talk. The footage is silent and overlaid with a commentary as the camera zooms in for a medium close-up. Parsons recalls how he noticed that 'she was wearing an unusual buckle on the shoulder of her gown'.[43] Sabrina poses and turns slowly, with the camera cutting in for a close-up of the celebrity in her satin dress (Figure 17). (In case we should miss the coveted accessory Parsons' hand is visible in shot, pointing to Sabrina's sparkling buckle). As the camera moves to a different area of the foyer the viewer is further informed of cosmetic details as Parsons explains how 'Sandra Dorne's earrings attracted much attention', while the elaborate and sparkling accessories are shown in close-up, in the foreground right of the frame. A similar approach characterised *Film Fanfare*'s wider reporting on the female star, as best exemplified by the footage of Dorothy Dandridge, as she arrived at London airport.[44] Parsons' commentary explains how she appeared 'wearing a very striking *cafe au lait* wool ensemble with all accessories matching - hat, bag, shoes, gloves, and umbrella'. The camera then cuts in for several close-ups as Parsons refers to each of the items in turn: 'here are the handbag and gloves... and the stiletto heeled shoes... the umbrella. Isn't she

Figure 17: At the Yield to the Night premiere: 'Sabrina was wearing an unusual buckle on the shoulder of her gown' (1956, edition 19)

lovely? Come on in for a close-up'. These shots are striking in their self-consciousness, and there is a pause between each of them which allows the viewer to contemplate the items in turn.

In terms of film premieres, general media coverage had long since foregrounded the attire of the female star, but television inflected this material with new possibilities. Coupled with the use of the close-up shot, the commentary seeks to be more 'intimate', picking up on details which a broader perspective might miss. (The male voice of Parsons makes for a rather strange intervention in this respect, as the coverage seems to assume a female viewer who is interested in the latest stylish accessories, particularly if they are worn by the stars). This coverage can be contextualised by noting that it was often the British stars who appeared in the fashion columns in women's and film fan magazines. In the case of *Picturegoer*, these columns advised female readers on how to wear the latest fashions, colours and accessories, or how to apply cosmetics or achieve a desirable posture and figure. (In the late 1950s two of the most regular columns were 'Let's be beautiful' by glamorous actress Arlene Dahl, or 'Design for dressing' by Barbara Taylor).

Emphasising the importance of the co-existence or interaction between these intertextual sites, Rank star Maureen Swanson was seen at the *Safari* premiere on *Film Fanfare* while appearing in *Picturegoer*'s 'Design for Dressing' column, specifically with regard to 'how to accessorise'.[45] It is certainly difficult to reconcile notions of a 'restrictive British femininity' with the premiere footage of British stars such as Janette Scott, Belinda Lee, and most obviously Diana Dors. Dors arrived at the *Yield to the Night* premiere in her powder-blue Cadillac and sparkling sequinned dress as Parsons eagerly (and suggestively) informed the viewer that 'suddenly the excitement was terrific... Diana gave the press what can only be termed as "an eyeful"'.[46]

> Stacey notes that the central role played by film stars in fashioning the ideal female image emphasises the 'contrast between contemporary consumer markets and those of the 1940s and 1950s. The significance of stars to spectators' knowledge of consumer fashion is reconstructed as particularly focused in the light of the expansion and diversification of fashion markets since that time'
>
> *(1994, p.197) (see also Epstein, 2000).*

While the role played by television at this time was new, it actually seems to appear on the brink of change, and in fact look back toward previous decades. If part of the distinctiveness of the early cinema programme was its immersion within the cinema as still essentially a *mass* medium, its interrelations with fashion and the female star seem to be part of this context. By the end of the 1950s *Picturegoer*'s fashion advice columns looked rather insular, if not a little dated, and as Stacey notes above, the diversification of fashion markets after this time diffused, rather than concentrated, the fashion influence of the female film star.

Based on the evidence of her respondents, Stacey argues that the close-up of the female body in film perhaps fostered intimate pleasures between female star and viewer in which 'clothes were the currency of a shared femininity' (p.198). She explains how:

> Close-up displays of parts of the female body may have functioned, not to alienate and objectify, but to produce a fascination which was remembered as a form of intimacy by female spectators. Thus, ironically, the very fetishism and fragmentation criticised by feminist theory seems to have a rather different meaning for spectators whose memories of such effects can be understood as a form of personalisation of the Hollywood star otherwise kept at a distance on the screen
>
> *(1994, p.206).*

Particularly given its repeated use of close-ups to foreground garments and accessories, the stars seen on television occupied a smaller, less distant screen which may have reinforced the intimate pleasures outlined here. At the same time, Stacey's work lends itself to the possibility that such fetishisation has important links with fan culture. While she is talking about the film text, her argument complements John Ellis' (1992) suggestion that the intertextual circulation of the star is incomplete and incoherent, prompting an infinite cycle of consumption. These possibilities are combined in the television coverage in that, while we see the full moving image of the star, particular aspects are fetishised by the camera's gaze. We are left with the impression of particular fragments of the stars - Sabrina's buckle, Sandra Dorne's earrings, Dorothy Dandridge's shoes, and this parallels the incomplete nature of the intertextual circulation of the star. For my purposes, this has a particular significance for the representation of the female star and their possible address to the female viewer. Characteristic of the wider cultural construction of the star, the coverage invokes both intimacy and distance - again structured through specific aspects of television's modes of address.

If we return to the letter in *Picturegoer* which highlighted the lack of accessibility and cultural visibility surrounding the premiere, it is possible that television made these events more visible than ever before. This throws light on a common conception that the British film industry had an uneasy relationship with glamour and showmanship. The live film premiere is a product of the early interaction between film and television and to a certain extent, it is specific to this time. It was designed to showcase not only the celebrities and stars, but the technological possibilities of television and its implications for the representation of film culture. By 1957 the filmed coverage of premieres had become routine. When discussing ABC's *Box Office* (1957-8) which succeeded *Film Fanfare*, a critic in *Kine Weekly* recalled how 'The programme included a visit to a premiere. I don't remember which one, but it doesn't really matter since most of them are alike'.[47] A letter in *Picturegoer* the same year insisted that 'TV's *Picture Parade* is becoming a conducted tour of premieres instead of... a parade of picture extracts'.[48] While expressing negative sentiments, these comments suggest how the period of experimentation was short-lived as the coverage rapidly became conventionalised. This was also true of television's look 'behind-the-scenes' of filmmaking which, in its own way, also played a central role in constructing television's 'image' of the cinema.

'Movies in the Making': Television Goes 'Behind-the-scenes'
In terms of the BBC's earlier cinema programme *Current Release*, the concept of 'behind-the-scenes' referred to the basic *techniques* of filmmaking - the demonstration of editing and dubbing for example. From the BBC's point of view, this may have been linked to the logic expounded by film critic Roger Manvell that learning about the 'work and skill that goes towards the making of a film' enables one to 'attain a deeper appreciation of the film as an art form' (1951, p.1). This clearly became less of a priority in the later programmes, and the concept of the

behind-the-scenes approach became associated with a wider perspective which focused on films in production. This included, for example, what happened on and off the set, how scenes were filmed, the locations used and so on. This, of course, provided a further opportunity for coverage of the stars, and the behind-the-scenes material contributed to the general 'showbusiness' focus of the programmes at this stage. The shift in what constituted behind-the-scenes coverage was also shaped by changing production strategies within the film industry. These included, for example, the increasing production of large-scale big budget pictures and filmmaking abroad (trends that were more entrenched by the late 1950s).

Film programmes today often include footage of 'action' on the set and features on 'the making of' a particular film are routine. Such coverage is a vital part of an intertextual process which primes the film text for consumption (Klinger, 1991, p.119), and television's view 'behind-the-scenes' adopted a promotional rhetoric from the start. In fact, it is precisely its commercial logic that is central in understanding the complexities of this coverage which can be productively explored through Barbara Klinger's conception of the relations between intertextuality and aesthetic commodification. Television has played a key role in the construction of this 'behind-the-scenes' perspective since it began and this section considers the institutional, industrial and aesthetic factors which surrounded its emergence.

In the discussion of *Current Release*, it was suggested that its 'behind-the-scenes' rhetoric offered a new perspective to viewers. There were few other texts - at least those likely to be consumed by the general public - in which the general filmgoer would have encountered this subject. The later 'behind-the-scenes' perspective (much of which was likely beyond the resources of the earlier programme), drew more widely on existing media forms. As with the focus on stars and premieres, television was not responsible for developing this coverage. It built upon models established by other media while adapting and re-inventing them within its own textual form. Nevertheless, it was the case that the fan magazine invested relatively little space in the discussion of this subject. *Picturegoer* spent less time reporting on films in production than it did covering those on release and interviewing their stars. Interviews often involved discussion of where the stars were working and on which films, but this was seldom explored in any detail. The magazine's 'Around the Studios' (and similar columns) which discussed both the American and British contexts, offered 'snippets' of news on the studios' production plans, reporting which stars were working where. A reporter would occasionally visit a film set, and the coverage of *Trapeze* (Carol Reed, 1956) in 1955 offers an example of how this area was approached. In this edition we are informed (rather unsurprisingly) that Gina Lollobrigida does not perform her own aerobatics, and we read about the injuries and accidents endured by the stand-ins. We are told that the director, Carol Reed, relentlessly insists on very dangerous aerial tricks and how Burt Lancaster 'discusses every shot with [Reed], and often shouts suggestions from his... trapeze'.[49] In an earlier edition of the magazine there is an article on the British adventure film *Highly Dangerous* (Roy Baker, 1950). Here, the reader is privy to slightly more technical information and is informed that in order to create the soundtrack for a crowd scene, nine different voices were 'blended together for the master track'.[50] The article goes on to explain that 'To complete your disillusionment [the] producer... reminded us that the Middle Eastern scenery was filmed at Aldershot and Denham'.[51] This reference to 'disillusionment' recalls the debates discussed in relation to *Current Release* - whether the audience wanted to see how films were produced, or whether it threatened to destroy their enjoyment.

One of the most important examples of behind-the-scenes coverage in *Picturegoer* in the 1950s can be found in 1952 when the magazine published a six-week serial 'This is filmmaking' on the production of the new Rank film *The Net* (Anthony Asquith,1953) at Pinewood studios. The writer, Donald Hunt was given permission 'to sit in on preliminary talks, script and casting conferences',[52] and the six-week series charted the progress of the film through the initial stages of casting and screenwriting until the production actually began. The very construction of this feature as a 'special' serial indicates that it was not a perspective regularly covered in detail. This is also suggested by its introduction, which implicitly suggests a lack of familiarity with the subject: 'Making a film is a long, fascinating and anxious business. Snippets about production appear from time to time in the newspapers... It all sounds very easy, but in fact, it's an expert and worrying job'.[53] While *Picturegoer* certainly reported on films before they reached the screen, this was often only in brief terms, and photographs of action on the set were relatively rare. Furthermore, as the decade progressed and the magazine increasingly pursued the teenage market, there was even less space for these issues as film jostled for room with pop music and television. By the time *Picture Parade* and *Film Fanfare* were offering a behind-the-scenes glimpse on a weekly basis, readers of *Picturegoer* were less likely to be reading about films in production than consulting the pop feature 'Disc Parade' (and it is difficult to imagine the series on *The Net* being included at all by this time).

Film programmes on radio perhaps offered the closest comparison with the coverage produced by television. For example, in the late 1940s, radio's *Picture Parade* featured 'Round the Studios' with reporter Roy Rich. Scripts indicate that in one edition he visited the set of *Nicholas Nickleby* (Alberto Cavalcanti, 1947) at Ealing studios and chatted to the production team.[54] On another occasion Rich visited the set of *Great Expectations* (David Lean, 1946) and spoke to the director David Lean and star John Mills. According to the *Radio Times,* an edition of *Filmtime* in 1952 featured 'a visit to the set of *The Sound Barrier*' (David Lean, 1952) at Shepperton studios, including interviews with the stars Ann Todd and Nigel Patrick and, again, director David Lean.[55] This was by way of an outside broadcast unit and similar items were not uncommon. While presenting *Film Fanfare*, Peter Noble also featured on *Movie-Go-Round* and presented the item 'Around the British Studios'. Similar to his 'gossip spot' on *Film Fanfare*, however (and the columns in *Picturegoer*), this was a very rapid round-up of production plans and stars' movements. Given that *Movie-Go-Round* began in 1956 and ran simultaneously with TV's *Picture Parade* and *Film Fanfare*, it is perhaps indicative of the extent to which television was beginning to colonise this area, taking over the role of radio.

One other possible source of the behind-the-scenes perspective is films themselves. Writing of self-reflexive films, a famous example of which is *Singin' in the Rain* (Gene Kelly/ Stanley Donen,1952), Jane Stokes argues that there 'is a pleasure in watching *films about films* and a thrill in seeing the means of production of a medium revealed [original emphasis]' (1999, p.3). Richard Maltby states, however, that *Singin' in the Rain* remystifies the process of production at the same time as it purports to demystify it (1995, p.48). That is not to imply that, in contrast, television offers a form of direct, unmediated and 'authentic' access. As with the coverage of stars, the behind-the-scenes material is constructed, selected and narrativised. Yet in the star interviews discussed, we are under the impression that we are seeing the stars 'outside' of their fictional roles, and the same is true of the behind-the-scenes coverage. If the cinema programme's construction of the star seeks to offer them 'as they really are', the behind-the-scenes footage attempts to offer the production process 'as it really is'. In fact, the affinity with the intertextual construction of the star is illustrated by Richard Dyer when he describes the star biography as an attempt to go 'behind-the-scenes of the star image' (1987, p.11). Although aspects of a behind-the-scenes approach had been developed by fan magazines and on radio, television offered new possibilities for this intertextual material. As we shall see, the behind-the-scenes perspective also became institutionalised at this time, when the film industry involved itself in its production in ways which had not occurred with radio.

Because of ABC TV's links with ABPC, it had easy access to Elstree studios, and it is not surprising that *Film Fanfare* was quick to adopt a 'behind-the-scenes' approach. Because of its affiliation with ABPC it seems that in comparison with *Picture Parade*, *Film Fanfare* included more coverage of the British studios. Existing editions include an item on the making of *The Baby and the Battleship* (Jay Lewis, 1956) at Shepperton studios, *My Wife's Family* (Gilbert Gunn, 1956)[56] at Elstree, Hammer's *X the Unknown* (Leslie Norman, 1956) at Bray studios[57] and several other examples. The behind-the-scenes material, of course, offered a further opportunity to promote the Associated British films. When watching *Film Fanfare*, the viewer heard of the pictures before they reached the studio floor (in Peter Noble's gossip spot), while they were in production (in the behind-the-scenes material), then finally when they were on release in John Fitzgerald's preview section. The 'behind-the-scenes' perspective is further evidence of how the institutional background of *Film Fanfare* shaped the textual content of the series. *Picture Parade* included its own coverage of film production in Britain, whether at the studios or on location. This covered a wide range of films such as *Dunkirk* (Leslie Norman, 1958), *Ice Cold in Alex* (J. Lee Thompson, 1958), *The Vikings* (Richard Fleischer, 1958) and *Next to No Time* (Henry Cornelius, 1958).[58] The making of the American films *Raintree County* (Edward Dmytryk, 1958)[59] and *The Inn of Sixth Happiness* (Mark Robson, 1958) were the subject of more than one edition. Produced by British MGM and British Twentieth Century Fox respectively, these were large-scale productions with big star casts, and their British production offered an opportunity for television coverage. Whether with regard to studio or location footage, the material shot by BBC and ABC generally focused on the British context, but coverage of Hollywood was certainly not absent. Both programmes used material from an American feature, *Warner Bros Presents* (1955-57) which is important here on a number of different levels. Firstly, it enables this study to offer more than a comparison with the American context by considering aspects of cross-cultural interaction: how the relations between film and television in America actively *shaped* those in Britain. Secondly, the Warner Bros material was hailed as a particularly successful example of behind-the-scenes coverage which was because it demonstrated how the specificities of television could best exploit this perspective. It also prompted a shift in the institutional relations between the media as it encouraged the British film industry to produce behind-the-scenes coverage themselves, marking a significant point in the media and industrial relations surrounding the cinema programme.

'Behind the cameras' at Warner Bros

As discussed in Chapter One, there were no programmes on American television that were comparable to *Current Release*, *Picture Parade* or *Film Fanfare* in the early 1950s, and variety programmes (such as *The Ed Sullivan Show*) offered the primary place for the coverage of film culture, a situation that was entrenched by the mid-1950s. While Ed Sullivan may have converted his programme into 'an hour-long advertisement for Hollywood' (Anderson, 1994, p.160), the most significant intervention in film programming came in 1952 with Walt Disney's successful series *Disneyland*. The programme essentially aimed to unite the Disney films, cartoons, products and amusement park into a vast commercial web, a consumer environment based upon a logic of 'total merchandising' (see Anderson, 1994, pp.141-155). By 1954 it was clear that broadcasting regulations would bar the US majors from having a significant stake in television, and that their only involvement with the medium would be as programme suppliers (Stokes, 1999, p.31). This included telefilm series and other forms of programming, but it was initially the possibilities opened up by *Disneyland* that provided an incentive for the companies to consider television production. In 1955 Warner Bros, Twentieth Century Fox and MGM all diversified into television with flagship series carrying the company name: *Warner Bros Presents*, *Twentieth Century Fox Hour* (1955-6) and *MGM Parade* (1955-6). Several studies refer to these programmes (Balio, 1990, Boddy, 1993, Anderson, 1992, 1994) and they are often described as an unsuccessful venture before the majors' institutional move into television production. Unlike *Picture Parade* and *Film Fanfare*, *Warner Bros Presents* was not a programme devoted to film, but a telefilm series which ended with a six- to eight-minute segment featuring Warner Bros films. Warner Bros were to produce three series of thirteen episodes, each loosely based on three of the company's successful films - *Casablanca* (Michael Curtiz, 1942), *Cheyenne* (Raoul Walsh, 1947) and *King's Row* (Sam Wood, 1942). Each week, the narrative episode of the series was followed by a section called 'Behind the Cameras at Warner Bros' and consisted of excerpts from a current film and appearances by its stars in what was described as 'exclusive behind-the-scenes production footage' (Anderson, 1994, p.169). According to Anderson, this section was Warner Bros' primary reason for producing the series and the network and sponsors complained about the 'over-commercialisation' of the programme and the 'incessant plugs' for the Warner Bros studio (Anderson, p.197). An article in the American magazine *Sponsor* objected to the promotional rhetoric of all flagship series, citing it as the primary reason for their low audience ratings (Boddy, 1993, p.148). After much debate, *Warner Bros Presents* appeared for a second series but under increasing pressure, Warner Bros reluctantly agreed to remove the 'Behind the Cameras' section. Anderson suggests that this marked the point at which the company 'redefined its role in television, becoming a movie studio that produced TV programming, rather than a [studio] which exploited television to publicize [sic] feature films' (Anderson, p.218). (The western series *Cheyenne* ultimately proved to be a considerable success and represented a template for Warner Bros' move into telefilm production).

It was the much criticised 'Behind the Cameras' section that was used in *Picture Parade* and *Film Fanfare*. As *Kine Weekly* observed, (British) Warner Bros 'are finding it quite an asset that that [its]... parent company [produces]... *Warner Bros Presents*.. The pre-filmed interview and trailer-type material... is ideal ready-made footage for the film magazines of either channel over here'.[60] Anderson suggests that in producing programmes which promoted current films, the major companies were left with texts containing no residual value. Blinded by their short-term plans for studio publicity, they could not recoup their losses through repeats or syndication (1994, p.212-3). While this may be so, Warner Bros ensured that they maximised the potential of the footage in other ways. With a keen eye on the British market, they saw an opportunity to recycle their material and a chance for free publicity. This was particularly so in the case of *Film Fanfare*, given that Warner Bros held substantial shares in ABPC and the Warner films were always shown on the ABC circuit. 'Behind the Cameras' was thus an excellent opportunity for exploitation. It not only provided material for ABC TV, but it simultaneously promoted the Warner films playing on the ABC cinema chain.

Anderson does not analyse the 'Behind the Cameras' section in detail and analysis is constructed from archival footage,[61] excerpts in the British programmes, and the British trade press. Like the American television network and sponsors, Anderson dismisses the 'Behind the Cameras' section as a cheap, unworthy 'plug', preferring to focus on the institutional negotiations and the telefilm series. Aside from its commercial motives, there is no suggestion as to the logic behind the section - its textual structures and modes of address - nor the perspective it offered to the 1950s television viewer. The discussion of the 'Behind the Cameras' section in Britain was very different, at least where the trade press coverage is concerned. The Hollywood industry was automatically perceived to be more sophisticated in the field of publicity, and *Kine Weekly* was particularly impressed by the material and described it as an example of 'typical Hollywood thoroughness in studying TV as an advertising medium'.[62] In the context of the trade press the footage was precisely admired, rather than denigrated, for its promotional logic. It was not simply that the British film industry were considering the material from a different perspective, as 'Behind the Cameras' was consumed outside of its original context. In America, it was not only the feature's 'overcommercialisation' that was considered problematic. The network and sponsors complained about its fragmented structure and abrupt textual shifts between the telefilm series, the advertisements, and the studio promotion (Anderson, 1994, p.195). In Britain, however, 'Behind the Cameras' was inserted into a different textual flow which altered its reception.

In *Picture Parade,* the American material could be found in the 'Hollywood Report' and in one edition, it focused on the John Ford western *The Searchers* (1956).[63] *The Searchers* was shot on location in Monument Valley, Utah and, combined with the large-scale nature of the production, this provided much of the interest of the coverage. As on-site activities at the Hollywood studios declined, the majors shifted into financing and distributing independent pictures, many of which were shot on location. *The Searchers* was produced by Warner Bros but as the coverage makes clear, on-location footage represented an important aspect of 'Behind the Cameras'. In fact, the behind-the-scenes coverage generally focused on interesting locations as a particular point of its appeal. This was often the case with exotic, foreign locations, the use of which was increasingly linked to the emergence of 'runaway' production as the decade progressed. While more prevalent in the 1960s, shooting abroad enabled the American companies to invest earnings blocked on domestic ground. In addition to providing authentic locales and perhaps European stars, foreign shooting offered the opportunity to hire cheap labour, and to benefit from European government legislation designed to protect domestic industries from US competition (Guback, 1985, p.478, Maltby, 1995, p.69). The behind-the-scenes coverage exploited industrial shifts occurring at this time, while it simultaneously existed as a product of these changes.

Warner Bros Presents was compèred by the American film actor Gig Young. In terms of *The Searchers*, he spends considerable time explaining how the crew moved into the 'brooding wilderness' of 'Navaho country, untouched by civilisation'. *The Searchers* was filmed at the largest Indian reservation in America and the coverage initially took a pseudo-ethnographic perspective, explaining the customs and lifestyle of the 'native' Navahos. In his introduction, Gig Young offers a sneak preview of what is to come and explains how 'You'll stand behind director John Ford as he gives star John Wayne his final direction for a key scene'. While visits to the set on radio often involved a discussion with the director, television added a visual dimension – the viewer could see the director at work. Gig Young then interviews the star Natalie Wood about her role in the film. Although the coverage aims to give the impression that this is taking place on location (with Natalie Wood in her Indian costume), it is evidently filmed in the studio. (In order to reinforce the impression that he is in the valley, Gig Young later drives off in a jeep). We then see some stunts performed and this is followed by coverage of the crew obtaining a difficult shot of horses galloping overhead for which it was

Figure 18: Gig Young and Jeffrey Hunter introduce the behind-the-scenes story of The Searchers (1956)

necessary to insert the camera into the ground. Again, this exploited the visual nature of television and this is further emphasised as a camera speeds along the dusty plain while Gig Young explains: 'And you'll be there, riding the camera car through the Indian village as the Searchers attack the wild Comanche'. Although shot on film, this drew upon the concept of a strong co-presence between viewer and screen associated with the live broadcast - the feeling of literally 'being there' amid the action.

The coverage explores the visual possibilities of television on a wider scale, and this is closely related to its promotional rhetoric, as first established in *Disneyland*. While its primary function was to advertise the increasing array of Disney products, *Disneyland* disguised its promotional ballyhoo by framing the programme with an *educational* discourse in which the behind-the-scenes material played a key role. As Anderson elaborates, 'Disney defined television as a companion medium to the cinema, as an informational medium which could be used to reveal the process of filmmaking... While Disney movies were presented as a seamless narrative, television gave [the] license to expose their seams' (1994, p.145). Before *Disneyland* began, Walt Disney had appeared on *Current Release* to discuss the process of animation, and the BBC screened a film (especially made for television) called *Operation Wonderland*, demonstrating Disney's craftsmen at work.[64] Anderson suggests, however, that:

> Disney's willingness to display the process of filmmaking suggests that reflexivity in itself is not a radical impulse. More a disciple of Barnum than Brecht, Disney had no intention of distancing his audience from the illusion in his movies. Instead, he appealed to the audience's fascination with cinematic trickery

(1994, p.145).

Disney exhibited what historian Neil Harris describes as an 'American vernacular tradition' as exemplified by Victorian showman P. T. Barnum. The logic of Barnum's showmanship depended on the assumption that the public:

> delights... in being fooled by a hoax and in discovering the mechanisms that make the hoax successful... Barnum encouraged an 'aesthetic of the operational, a delight in observing process and examining for literal truth'.[65] Far from being hoodwinked by Barnum's artifice, the audiences that witnessed his exhibitions took pleasure in uncovering the process by which these hoaxes were perpetrated

(Anderson, 1994, p.145).

Anderson suggests, therefore, that Disney created his own form of 'operational aesthetic' centred on the process of filmmaking and animation. With its ability to make people and animals come to life, the art of magic and illusion has particular links with animation, yet this logic can also be applied to the construction of 'Behind the Cameras'. Warner Bros adopted their own kind of operational aesthetic and incorporated this into the coverage of their films. For example, *Kine Weekly* described the example of *The Gun Runner* (Gordon Douglas, 1956) (a Warner Bros action thriller in which Alan Ladd sails to Cuba in a paddle boat with a consignment of guns). *Picture Parade*'s 'Hollywood Report' demonstrated how the jungle scenes were created by the props department, then displaying the finished effect in an extract from the film. With respect to *The Searchers*, we are shown how a daring cliff stunt is performed and as we see the final cut in the film we are told: 'And this is how you'll see it in *The Searchers*'. If we apply P. T. Barnum's logic to this coverage, the final scene is effectively the 'hoax': Jeffrey Hunter's character is not dangling dangerously from the rugged cliff top (nor is it Jeffrey Hunter), just as Alan Ladd's 'exciting experience' in the jungle is really in the Warner Bros studio. The point of the coverage, however, is not to expose the illusory nature of film but to investigate the *how* - how the visual illusion - the 'hoax', is actually performed.

The contextual significance of such material can be productively explored through Barbara Klinger's theory of the cultural relations between a film and its intertexts. Klinger suggests that certain types of textual response are motivated by intertexts which include, for example, posters, trailers, media coverage and merchandising. These forms (what Stephen Heath refers to as 'epiphenomena') define the text as a product:

> Such promotional forms... exemplify a relation between intertextuality and aesthetic commodification; they operate as an intertextual network designed to identify a film for consumption. Films circulate as products, not in a semantic vacuum, but in a mass cultural environment teeming with related commercial significations. Epiphenomena constitute this adjacent territory, they create not only a commercial life-support system for a film, but also a socially meaningful network of relations around it which enter into reception

(Klinger, 1991, p.119).

As Klinger notes, this network has been intensified historically with the invention of new technologies which advance the cultural range of promotion (p.124). As part of this intertextual framework Klinger significantly includes the 'behind-the-scenes' view on the making of a film which, she notes, 'has a staunch place in the film promotion' (p.128). The signifying relation between the text and forms of promotional intertextuality is secured by placing 'textual elements within other narratives' (p.129). For example, a star interview relating to a contemporary film may frame the material with a 'story' on the star's career or personal life.

This intertextual narrativisation is particularly evident in television's behind-the-scenes material. In the coverage of *The Searchers* Jeffrey Hunter joins Gig Young to introduce the feature and explains: 'When John Ford was directing [the film]... he put a camera to work behind the camera. He got a record... of everything that happened from the time the ground was broken... to the time the last scene was wrapped up'. Implying that this was something of an innovation, Hunter refers to the material as 'a *diary* of a motion picture, the first of its kind as far as I know'. Proudly displaying a steep pile of film cans, Gig Young describes the footage as 'an entire picture about the making of a ... picture' and subordinating the film text to its television preview he promises: 'You'll relive an even larger story' (Figure 18). Given the film industry's attempt at this time to differentiate films from television through innovations such CinemaScope - in short to make them 'bigger' - this description of 'an even larger story' is intriguing. The emphasis on size here perhaps referred to the extended perspective of the coverage (what occurred 'behind the camera' as well as in front), but also to one of the primary emphases of the material - its location. This is not only so in terms of the footage of moving the crew into the valley and the problems this posed, but also the 'ethnographic' perspective on the native Navahos which functioned to establish the authenticity of the environment. As Klinger notes, the foregrounding of textual elements in media coverage often surpasses their intrinsic narrative function, steering reception toward particular aspects or themes (1991, p.124). This is comparable to the construction of the star image, only here it relates to an 'image' constructed for the film. This can be applied to the coverage of *The Searchers*. The setting is of course also a spectacle in the film itself, and *Variety* described it a 'VistaVision-Technicolor... excursion through the Southwest' which compensated for what it in fact perceived to be an overly repetitious and simple story.[66] Yet in the behind-the-scenes footage the location dominates the material, and it is explicitly foregrounded to provide the context for what is almost a 'boy's own' adventure story in the dusty 'brooding wilderness'.

According to Anderson, *Disneyland* offered the viewer 'an incentive to see the completed movie, because the movie itself provided the resolution to the story of the filmmaking process' (Anderson, 1994, p.145). In the coverage of *The Searchers*, the behind-the-scenes footage is punctuated by excerpts from the film which, as Gig Young repeatedly reminds us, is 'now playing in VistaVision and Technicolor at your local theatre'. However, argument suggests a rather simplistic relation between text and intertext. It not only presupposes a closed film text, but argues that this acts as the closure to the enigma created by the intertext, in this case, the behind-the-scenes coverage. As Barbara Klinger argues, 'The goal of promotion is to produce multiple avenues of access to the text that will make the film resonate as extensively as possible in the social sphere' (1991, p.124). This semiotic 'raiding' may not create a coherent reading of the film as the viewing process may be structured by digressions - what Klinger refers to as 'momentary guided exits from the text' (1991, p.129). This applies to a range of textual elements but in the case of the behind-the-scenes footage, the viewer may relate to the film through a '"how this was done" mentality shaped by their fascination from the exposé' (Klinger, p.131). (With the increasing use of sophisticated special effects, this is perhaps more prevalent in recent years, but is still applicable to the 1950s). Klinger's argument is based upon a hypothetical model of the spectator rather than historical and empirical audience research. Nevertheless it offers a way into considering how television's increasing emphasis on the behind-the-scenes coverage may have influenced the reception of film. It offered a framework of interpretation which primarily entered cultural consciousness with the advent of television in the 1950s. Klinger's argument does not prioritise text over intertext, and suggests a more complex relation between the two. However, when considering her argument from a historical perspective and applying it to the coverage in question, it is important to acknowledge that the film text may not have been part of the equation at all. Given that cinemagoing went into precipitous decline at this time, it is possible that some viewers may not have seen *The Searchers*, and their primary experience was the 'even larger story' 'behind-the-scenes', shown on television.

If the Warner Bros coverage exemplified what *Kine Weekly* described as 'typical Hollywood thoroughness in studying TV as an advertising medium',[67] this was because it exploited the visual nature of television to particular effect, as well as increasing the scale of the coverage beyond the means of other media. This is particularly so in terms of its bid to deconstruct how the film illusion is created (the 'hoax'). It may seem perplexing, then, that it received praise, rather than criticism from the trade press, for as established with respect to *Current Release*, divulging the 'secrets' of production was seen as bad commercial logic. Nor did the film industry change their views on this matter as the decade progressed. In 1959 complaints were directed at the BBC's schools programme *Looking at Films* because it intended to demonstrate the 'behind the scenes tricks of filmmaking'. In considering the positive response to the American material, a possible difference is that it was produced by the film industry, and was a form of promotion over which they exercised control. Yet the situation was more complex than this. The film industry's dislike of the behind-the-scenes aspects of *Current Release* was perhaps because the filmmaking techniques were demonstrated as an end in themselves, thus lacking the commercial logic outlined. The material was not organised around a particular film and as a result, it was not part of an intertextual narrative which served as an 'aperitif' to a seat in the cinema. Indeed, the trade press described these educational aspects of the programme as antithetical to its publicity function. As Klinger notes, however, when in the service of promotion, 'The film industry is more than willing to be self-reflexive; exposing the star behind the character, the mechanics behind a special-effects scene... [as it] ... provide[s] yet another means for fetishizing [sic] the text in question' (1991, p.131), that is, creating multiple avenues for publicity. In fact, the Rank Organisation was so impressed with the benefits of the behind-the-scenes perspective that it produced a trailer for cinema exhibition which let 'the public into the secret of what is going on at Pinewood studios'.[68] With the glamorous British star Belinda Lee as its guide, the audience was invited on a tour of the studio to witness the making of the films in production. This offers, then, an interesting example of television shaping wider exhibition practices.

As established, *Picture Parade* and *Film Fanfare* focused on films in production, rather than the techniques of filmmaking (such as editing and dubbing). While the material produced by the BBC and ABC was not necessarily criticised by the trade press (as with *Current Release*), it did not receive the praise directed at the American footage, indicating differences in approach. In terms of *Film Fanfare*, in his role as the 'roving reporter', it was John Parsons who presented the coverage of the British studios. In one edition Paul Carpenter introduced Parsons' spot by explaining, 'And now here's our on-the-spot reporter with another behind-the-scenes glimpse of movies in the making'.[69] It was not, however, a glimpse of 'movies in the making' in the manner of 'Behind the Cameras'.

Parsons' commentary accompanied silent film as he explained how he 'visited the various stages at Elstree to see who was there. I started on stage one...',[70] and as this suggests, the structure of the item was more of a *leisurely* tour. Parsons was simply dropping in to observe the daily activity of the studios, the purpose of which was less clearly defined. On this occasion he met Zena Marshall, Tony Martin and Vera Allen, who were starring in the new Associated British musical comedy, *Let's Be Happy* (Henry Levin, 1957). Once outside he encountered Joan Collins and crew on a tea break - at the studio to film the drama *Sea Wyf* (Bob McNaught, 1957) - followed by an encounter with the 'enchanting Janette Scott' who, with her 'fresh charm and unaffected beauty' was leaving for home after a day on the set of *The Good Companions* (J. Lee Thompson, 1957). This was the typical format: a leisurely chat with whichever stars happened to be around. It was another facet of the programme's star-orientated focus and as a result, the material was less organised around the *process* of filmmaking or a particular film. That is not to suggest that the Warner Bros material ignored the stars. The American television network ABC had commissioned the series partly because of the promise of 'glamorous star value' (Anderson, 1994, p.171). Yet the stars were integrated into the narrative process and their appearances were designed to forward this logic.

In *Film Fanfare*, the viewer was still 'privy' to glimpses of films in production, how scenes are shot, sets constructed and so forth, but there was likely to be as much emphasis on the tea break in between takes, than on the filmmaking process itself. This could also differ from the slick, commercially orientated material seen in *Warner Bros Presents*. *Film Fanfare* featured footage of the making of *Assignment*

Redhead (Maclean Rogers, 1956),[71] a British thriller in which a master criminal (Ronald Adam) is embroiled in murder as he flies to London from post-war Berlin. Although beginning with a shot of the set (Figure 19), the coverage mainly focused on Ronald Adam and Richard Denning as they played out a tense scene in the studio. For the majority of the take, there was little to distinguish it from viewing the scene in the final film. In this particular example, we do not see the scene direction or the cameras rolling. Unlike the operational aesthetic in *The Searchers*, the viewer is not privy to the construction of any particular stunts or shots, nor is it interwoven seamlessly with excerpts from the final film. Of course, a film such as *Assignment Redhead* does not offer the same possibilities for coverage as a Hollywood picture like *The Searchers*, and this needs to be considered in terms of the more general coverage of British film production and the context of its limited resources. Nevertheless, *Film Fanfare*'s coverage was less adept at exploiting the visual possibilities of this material. While we are obviously given a visual perspective, there is little to distinguish the format from that previously used on radio. Furthermore, in contrast to Peter Noble's 'gossip spot' in which the news was presented as though it came from 'inside' the film industry, Parsons was more of a distanced observer. When he introduced the coverage of the star previously discussed ('as they really are, and in close-up'), Parsons had explained to the viewer how his job was 'to represent you, and together we shall go out to airports, docks, stations and meet the stars as they arrive...', and this is an accurate description of Parsons' role more generally. He functioned to stand *in* for the audience, constructing a space from which they could observe the action. Despite the programme's close industrial links with ABPC, Parsons' commentary foregrounded his presence at the studios as a visit, which implicitly suggests an outside perspective (perhaps conflicting slightly with Noble's utterly 'inside' perspective). Although on certain occasions he spoke directly to the camera from the film set or location, many visits were on silent film with an accompanying commentary. This was similar to much of the coverage of premieres and stars and contributed to the distance outlined. (This raises the interesting point that while television's behind-the-scenes coverage offered a visual perspective, it could sometimes loose the *aural* dimension offered by radio). Rather than attempt to divulge secrets from 'inside' the film industry, Parsons offered a commentary on the activity. This was in contrast to 'Behind the Cameras' which, as Christopher Anderson notes, offered 'an "insider's view" of studio productions' (1994, p.192).

In *A Seat at the Cinema*, which revealed 'the "inside" story of filmmaking... [and] the parts played by the many technicians and specialists' (1951, p.1/142), Roger Manvell spoke of the need to 'strip away the spurious glamour' of the film in order to understand this complex process. Whether in terms of the American or British coverage, the behind-the-scenes material certainly tended to glamorise as much as reveal, and it offered a very selective perspective on filmmaking. The work of 'many technicians and specialists' was subordinated to more visible and aesthetically

Figure 19: Film Fanfare goes 'behind-the-scenes' on the set of British thriller Assignment Redhead (1956, edition 9)

pleasing technical accomplishments. We see nothing, for example, of the labour relations or economics of filmmaking, the long hours or hundreds of retakes. Rather, we are presented with the talents of stars and directors, the rolling cameras and the stunts and sets.

'Business is Booming': Constructing an Image of Film Production

As this suggests, this increased visibility of films in production, and the entire concept of going 'behind-the-scenes' circulated particular constructions of the film industry which are an important part of its historical significance. The traditional Hollywood studio system was in decline by the late 1950s and was undergoing a period of transition, yet this is not the impression given by the Warner Bros material. Firstly, in terms of its use in Britain, both programmes claimed Gig Young as their own personal link with the dynamic world of Hollywood. *Picture Parade* would announce that it was time 'to hand over to our Hollywood correspondent' while *Film Fanfare* would explain how 'Gig Young is waiting for us in the Burbank studios, California', and the use of inclusive terms ('our', 'us') was an attempt to anchor his address to the British viewer. In each programme, the footage actually came from *Warner Bros Presents*. Given that *Picture Parade* and

Film Fanfare ran simultaneously, they may have avoided using the same extracts, as duplication would clearly have compromised their claims to exclusivity. In general, the footage presented an image of a bustling, *thriving* film factory, a regime which, paradoxically, television was helping to render obsolete.

In one edition *Picture Parade*'s 'Hollywood Report' focused on the epic western *Giant* (George Stevens, 1956) which represented 90% of Warner Bros' profits in 1956 - a fact which indicates how the previous studio-based production regime of moderate features was in decline (Anderson, 1994, p.219). As Anderson suggests, however, like Disney, Warner Bros planned to use television to revive their image 'as a traditional Hollywood studio, an identity far more marketable than its new identity as a distributor of independently produced movies' (1994, p.192). To a certain extent, a similar impression was given of the British film industry. This was a time when two of the main British studios, Pinewood and Elstree, were experiencing considerable difficulties (Macnab, 2000, p.203). In 1957 the Rank Organisation dropped the contracts of many of its key stars because it could no longer afford them. ABPC also made cutbacks, and production at Elstree was halved (Ibid). However, it is difficult to reconcile this situation with media coverage at this time, particularly that produced by television, as well as other intertexts such as *Picturegoer*. While happy to report on the declining fortunes of the Hollywood industry, *Picturegoer* usually presented the British film industry in a more positive light. In 1958 Elstree is still constructed as home of the Hollywood stars and a veritable international film centre, and in its 'Pinewood Souvenir' issue in 1957 (to celebrate the twenty-first birthday of the studio), we are told that 'All... [the studio's] stages are busy. The huge ... restaurant buzzes every lunchtime with ideas for future films, ... schedules [are] going right through 1959'.[72] The discourse in *Picturegoer* complemented that of television, particularly Peter Noble's 'gossip spot' in *Film Fanfare*. As Noble explained in one edition, 'We're so full up in British studios... that we've had to send a film unit off to Rome'. This was with reference to the film *The Little Hut* (Mark Robson, 1957) and Noble proudly reiterated that 'We are making British pictures in every studio *choc-a-bloc*, and that's excellent news for us all'.[73] This worked in dialogue with the image constructed by the behind-the-scenes coverage which, like the material on Hollywood, presented a thriving, self-enclosed enterprise (particularly at Elstree studios). This is despite the fact that, as David Quinlan has written, 'The days of the big studio in Britain, financed by its own proprietor's films' was coming to an end (quoted in Macnab, p.203). By the late 1950s, rather than a studio like the Rank Organisation financing their own quota of films, the emphasis shifted to co-productions. Stars were as likely to be attached to individual producers as big studios, and as in Hollywood, they were more inclined to form their own production companies (Macnab, 2000, p.203). However, if it was Howard Thomas' aim to use ABC television to improve cinema admissions, this would hardly have been fulfilled by reporting on ABPC's cut backs, just as it would have been bad commercial logic for *Picturegoer* to persistently dwell on the problems of 'British filmdom'. *Film Fanfare* had a particular interest in promoting an image of a bustling, healthy British film industry - of which ABPC was of course part. This is in some ways similar to the construction of the star and television's careful regulation of knowledge relating to their off-screen lives. A discourse which might conflict with the programmes' celebratory, affectionate and admiring attitude toward the cinema is generally suppressed in favour of a more positive, reassuring (and of course entertaining) perspective. In a similar manner to the interviews, we are presented with a narrative which interacts with, and contributes to, the wider narrative of film culture constructed by the programmes at this time.

It is significant that the American material influenced the relations between the film industry and television in Britain, as certain companies moved into the production of 'behind-the-scenes' material themselves. Following the lead of Warner Bros, certain British-based companies began to produce behind-the-scenes material themselves, shooting the extra footage while they were making the film. While the detail of the individual examples cannot be discussed in detail here - not least because they no longer exist - this move included companies such as Warwick,[74] British Twentieth Century Fox and the Rank Organisation. The material was not, however, shown as part of *Picture Parade* and *Film Fanfare*. Evidently noticing the benefits of coverage organised around a particular film, the film companies produced material for one-off programmes. Between fifteen and thirty minutes in length, these were broadcast by the ITV companies on an occasional basis, although usually late on a Saturday night. Following the lead of the American material, several of the programmes derived much of their interest from the location of the production. The Fox programmes focused on the all-star features *Anastasia* (Anatole Litvak, 1956) (shot in Paris), *Island in the Sun* (Robert Rossen, 1957) (filmed in Grenada and Barbados) and *Heaven Knows, Mr. Allison* (shot on a Pacific island) (John Huston, 1957). The Anglo-American Warwick had a particular penchant for exotic adventure films and programmes were organised around *Zarak* (Terence Young, 1957) (filmed in Morocco), *Safari* (filmed in an unidentified 'jungle' location) and *Interpol* (shot in Italy). The Rank features were less exotic and focused on the films *Hell Drivers* (C. Baker Endfield, 1957) (concerning the exploits of a group of rival lorry drivers, including Stanley Baker, Patrick McGoohan and Herbert Lom) and *Miracle in Soho* (Julian Amyes, 1957) (in which a Soho roadworker [John Gregson] falls in love with a barmaid [Belinda Lee]). While *Hell Drivers* was shot on-location, *Miracle in Soho* was a studio production. Although it is unclear whether they used an 'operational aesthetic' like 'Behind the Cameras', the *TV Times* clearly constructed these features as an intertextual narrative. They were often accompanied by a feature article, and in the case of the *Anastasia* programme the viewer is promised 'an account of how... [the] film was made, the story of what went on behind the cameras as well as in front'.[75] In terms of *Island in the Sun* the *TV Times* boasted that 'Seldom is the story behind the making of a film told in film form itself...'[76] However, rather than representing an innovation, this attested to the increasing *conventionalisation* of this narrative approach.

Although the trade press attributed the production of such features solely to the film companies (claiming, for example, that companies such as Warwick and Rank had set up special 'TV units'),[77] it is clear from the *TV Times* that the situation was more complicated than this. The billing of the programmes attributed both script and direction to television personnel, and it would seem that although the film companies did shoot the extra material while the film was in production, the ITV companies were responsible for incorporating it into a programme - editing the material, writing a script and providing a presenter. In fact, in contrast to the trade press which attributed the programmes to the *film* companies, the *TV Times* implied that they were solely produced by television.

As far as television was concerned, there may have been issues about advertising. While the regular cinema programmes had tested the boundaries in this respect, these other features were perceived as more problematic. In August 1957 the *Daily Film Renter* published a revealing announcement by Associated-Rediffusion (AR) which explained that:

> We have learnt that these film industry pictures could be construed as sponsored entertainment and come into conflict with the terms of the Television Act. Rather than get into hot water, we have decided to stop showing them as soon as we have wound up our existing contracts with film companies.[78]

AR's statement demonstrates an acute awareness of the regulation of 'promotion' on television at this time which, as discussed in relation to the advent of ITV, was a particularly hot topic of debate. The television companies thus had a vested interest in stressing their editorial control over the programmes. With regard to the film industry, the bid to claim responsibility for the features was perhaps an attempt to exercise a discursive control over television. To be involved in television production was to have a stake in the new medium, to take matters into their own hands, and this was an attractive prospect to the industry faced with an increasingly uncertain future. It can equally be seen as a more literal bid for power. Much of the tension surrounding the cinema programme (ranging from the issue of 'criticism' to that of scheduling) seems to have been related to the industry's gradual loss of control over a centralised sphere of *exhibition* - films could now be viewed and previewed elsewhere. The moves made by the film companies to produce television material themselves can be understood as a strategy for diversification (in terms of the move into television programming more generally, at least with regard to the Hollywood majors and Warwick). Warwick were also seeking to make a deal to produce telefilm series at this time based on a number of its successful feature films. But it can also be perceived as an attempt to reassert control at the level of production.

While only traces of the exchange exist, evidence suggests that coverage of film culture in the 1950s - particularly behind-the-scenes features, moved across national borders quite freely. This traffic appears to have been largely one-way, with Hollywood influencing the British industry, but it complicates the notion that there were clear contrasts in their situations. The British film industry were no less eager to experiment with the possibilities of television, they were possibly just less sophisticated in their approach. Most importantly, the institutional structure of British broadcasting militated against more extensive interaction. The examples of Warwick, Fox and Rank indicate that, to a certain extent, there was an attempt to duplicate the relations between American film and television, a situation which enabled film companies (although barred from owning a stake in the medium) to produce programmes promoting the studio's own wares. Programmes on 'the making of' a specific film are routinely broadcast on television today, but in the late 1950s institutional and cultural factors made this more problematic. It remains, however, an area in which film companies produce footage themselves, offered as part of the pre-prepared publicity for a film (Kerr, 1996, p.141), so the strategies developed in the 1950s have a wider historical and industrial significance.

The different types of coverage discussed in this chapter encapsulate the cinema programme's approach to the cinema - exuberant, glamorous, affectionate and admiring - and above all, often exciting and new. A product of the close relations between the media, the cinema programme was involved in constructing a particular image of British film culture. But, according to conventional accounts, this was also a time when, in other respects, the cinema was aiming to resolutely *differentiate* itself from television - as based around innovations in technology (colour, widescreen) and film content (narratives and images which pushed the boundaries of film censorship). As a connecting textual, technological and cultural space between cinema and television in the 1950s, the cinema programme complicates these claims to autonomy and competition. This relationship is the subject of the next chapter.

Notes

1. 11 August, 1954. Alan Sleath to Cecil Madden. T6/177.

2. 1 July, 1954. Cecil Madden to Daphne Turrell. T6/177. The programme was produced by Alan Sleath who would later produce *Picture Parade*.

3. As reported by *Kine Weekly*, 15 December, 1955, p.167.

4. See BFI microfiche on *Carousel*.

5. *Kine Weekly*, 1956 Box Office Poll, 13 December, 1956, p.7.

6. BFI microfiche on *Carousel*.

7. *Daily Sketch*, 17 April, 1956. BFI microfiche on *Carousel*.

8. *Picture Parade* script, no.2. Transmitted 30 April, 1956.

9. 'First Television Film Premiere'. T6/399/1.

10. *Picture Parade* script, no.2.

11. 'First Television Film Premiere'. T6/399/1.

12. *Evening News*, 28 August, 1956. BBC press cuttings box, PP62.

13. 'BBC's Roving Eye Goes to Premiere', undated TCF press release. T6/403/1.

14. As reported in *Motion Picture Herald* (27 December, 1952, p.1).

15. *Motion Picture Herald*, 27 December, 1952, p.1.

16. *Daily Film Renter*, 16 April, 1956, p.3.

17. Peter Dimmock, 'Television Out and About' in *BBC Yearbook* (1951) p.51.

18. *Radio Times*, 18 January, 1952, p.44.

19. *Daily Express*, 30 April, 1956, WAC press cuttings, 1955-58. Such comments continued well into the early 1960s. See ITC press cuttings on the screening of feature films.

20. *Kine Weekly*, 4 April, 1956, p.5.

21. These were produced by Associated British Pathé so it is not surprising that they quickly became a regular feature in *Film Fanfare*. Although less extensive, the newsreel coverage was to a certain extent similar to the filmed premieres in both programmes and it would seem that ABPC recycled the material for use in both media. Television was taking over the role of the cinema newsreel more generally at this time, and the broadcasting of the film premiere can be seen as part of this process.

22. *Picturegoer*, 10 April, 1954, p.3.

23. *Kine Weekly*, 13 December, 1956, p.6.

24. *Variety*, 4 July, 1956, p.55.

25. *Variety*, 4 July, 1956, p.55.

26. *Picture Parade* script 7b.

27. *Daily Herald*, 14 September, 1956. BFI microfiche on *The King and I*.

28. Fox press release: 'BBC's "Roving Eye" Goes to a Premiere'. T6/403/1.

29. *Picture Parade* script 23a.

30. *Daily Film Renter*, 12 August, 1956, pp.22-3.

31. *Radio Times*, 11 September, 1955, p.12.

32. *Today's Cinema*, 12 April, 1956, p.2.

33. *Kine Weekly*, 19 April, 1956, p.22.

34. *Daily Film Renter*, 14 September, 1956, p.6.

35. The Eurovision link came into regular use in 1955 and enabled the same programme to be transmitted simultaneously in several different countries. It also made possible the exchange of programmes between participating countries including Britain, France, Western Germany and Switzerland. As *Picture Parade* suggests, it could also be used to link up to another country for an interview or event. Much more so than ITV, it was the BBC who were directly involved with the development of the Eurovision (Briggs, 1995, p.142). With regard to the *Italian Film Festival,* it seems that the Eurovision link was used to transmit the programme in Italy (1 July, 1954. Madden to Turrell, T6/177).

36. *Picture Parade* script, 14a.

37. *Film Fanfare* no.10. April, 1956.

38. 'Premieres - Are they Worth It?', *Picturegoer*, 14 February, 1956, p.14.

39. As above, p.15.

40. Scholars have investigated particular case studies which explore how Hollywood cinema, particularly between the 1920s and 1940s, has historically constructed the female viewer as consumer. See, for example, Allen (1980), La Place (1987), Doane (1989) and Gaines (1989).

41. xli *Film Fanfare* editions nos. 10 (*Safari* premiere) and 19 (*Yield to the Night* premiere).

42. *Daily Film Renter*, 14 September, 1956, p.6.

43. *Film Fanfare* no.19.

44. *Film Fanfare* no.9. Originally known as a singer, Dandridge rose to fame with the title role in *Carmen Jones* (Otto Preminger, 1954) and according to Karen Alexander, Dandridge holds the title of being 'America's first sex goddess of colour' (Alexander, 1991, p.47). In the television coverage she is constructed less as a sex symbol than the ultimate model of fashionable sophistication, but it nevertheless pivots on the implication that black could be 'beautiful'.

45. *Picturegoer*, 18 February, 1956, p.26.

46. *Film Fanfare* no.19.

47. *Kine Weekly*, 15 August, 1957, p.32.

48. *Picturegoer*, 14 September, 1957, p.3.

49. *Picturegoer*, 10 December, 1955, p.12.

50. *Picturegoer*, 6 November, 1951, p.5.

51. As above.

52. *Picturegoer*, 7 June, 1952, p.8.

53. As above.

54. liv *Picture Parade* radio script no.4.

55. *Radio Times*, 2 January, 1952, p.44.

56. Both in *Film Fanfare* no.1.

57. *Film Fanfare* no.2.

58. *Picture Parade* scripts 44b, 59b, 58b and 65b respectively.

59. *Picture Parade* scripts 47b and 52b.

60. *Kine Weekly*, 19 April, 1956.

61. The 'special collectors' edition of *The Searchers* (1956) on video contains footage from the coverage of the film on *Warner Bros Presents*. Unless otherwise indicated, all analysis of this footage is based on the video material.

62. *Kine Weekly*, 13 December, 1956, p.141.

63. *Picture Parade* script 17a.

64. *Current Release* script no.4.

65. Anderson is quoting here from Neil Harris' book *Humbug: The Art of P.T. Barnum* (Chicago, University of Chicago Press, 1973) p.79.

66. *Variety*, 14 March, 1956, p.55.

67. *Kine Weekly*, 13 December, 1956, p.141.

68. *Kine Weekly*, 2 May, 1957, p.22.

69. *Film Fanfare*, no.29.

70. *Film Fanfare*, no.29.

71. *Film Fanfare*, no.9.

72. 'Pinewood's got plenty to shout about', *Picturegoer*, 28 September, 1957, p.7. In the same issue

see also 'The starry build-up stars' (p.8-9) and for Elstree see 'This is Britain's Hollywood', 29 March, 1958, p.3.

73. *Film Fanfare*, no.24.

74. Founded by Irving Allen and Cubby Broccoli, Warwick was one of several British-based independent companies in the late 1950s which made their films in collaboration with the big American distributors, in this case, Columbia. Robert Murphy describes how Broccoli and Irving 'successfully instilled a Hollywood-like-lavishness into their ... film productions' (1992, p.219), and the majority of their features at this time were shot in Technicolor and/ or CinemaScope. They often used a mixture of imported leads (such as Alan Ladd, Victor Mature and Anita Eckberg) as well as recurrent British actors (including Nigel Patrick and Anthony Newley).

75. *TV Times*, 9 February, 1957, p.33.

76. 'ITV goes on location - to an *Island in the Sun*', *TV Times*, 19 July, 1957, p.29.

77. *Daily Film Renter*, 16 August, 1957, p.4.

78. *Daily Film Renter*, 16 August, 1957, p.4.

Chapter Seven

Looking at the Wider Picture on the Small Screen: Reconsidering Television, Widescreen and the 'X' Certificate in the 1950s

It is often suggested that one of the film industry's 'responses' to television in the 1950s was to 'give [the audience] something television can't' (Balio, 1985, p.23). Thomas Doherty, for example, offers a succinct summary of this view when he claims that 'the nature of television - its physical shape, episodic format, and family-orientated content - largely determined the nature of the motion pictures in the 1950s' (1988, p.25). This has come to mean essentially two things. Firstly, that films distinguished themselves from television in terms of *form* by offering widescreen and spectacle, and secondly that they offered innovations in film *content*. As John Hill claims, for example, in producing images of 'horror and sex', the 'X [certificate] film was a key weapon in differentiating cinema from television, allowing representations unlikely to be seen in the home' (1986, p.50). Yet a key premise of this book is that the cinema could equally not afford to ignore the promotional possibilities of the small screen. Economic imperatives created a situation in which the bid to distinguish film from television - to construct a 'different exchange value' for the cinematic product (Anderson, 1991, p.87) - was also a strategy in part *dependent* on television for its circulation. In fact, this is surely just as logical as the argument concerning film's differentiation from television. If widescreen, colour or shifts in film content were to lure audiences *back* to the cinema, then it was necessary to 'pursue' them wherever they had 'gone'. This meant addressing the audience in the domestic sphere through what was rapidly becoming the dominant site for the consumption of screen entertainment: the television. Thus, the explanation that new cinematic innovations sought to 'lure people from their TV sets' (Balio, 1990, p.23) is in many ways entirely correct, only we also have to acknowledge that this may have happened quite literally: that television played an *active* role in this process.

This chapter first explores the ways in which the cinema programme occupied a contradictory status in the bid to 'redefine' the aesthetic and technological identity of the cinematic product - both actively discussing the new technologies, while effectively accentuating the differences between the media. Focusing primarily on the X certificate film, it then considers television's role in discussing, circulating and displaying the shifting boundaries of film censorship. This suggests that innovations in film content and the regulation of film on television in fact developed *together*, each contributing to the institutional policies and cultural reception of the other. At a time when the cinema was deliberately negotiating a boundary struggle in its cultural role as screen entertainment, how did television play a role in circulating this new identity? This involves going back to the BBC's first regular TV cinema programme, *Current Release*, as well as looking forward to the latter period of *Picture Parade*'s development in the early 1960s.

Aesthetic and technological comparisons between film and television have historically favoured the cinema, constructing a dichotomy based on essentialist claims about the media (McLoone, 1996, p.81). This is particularly so when it comes to the 1950s, where the established historical narrative pivots on a technological contrast between the big screen and the small. The interaction between the film and television industries since this time, primarily from the 1960s onwards, has of course also lead to the screening of widescreen films *on* television. Fuelling claims about the aesthetic 'incompatibility' of the media, this has largely confirmed perceptions and prejudices surrounding television's longstanding 'deficiencies as an exhibition site for film' (Klinger, 1998b, p. 4). Yet with the advent of digital technology and widescreen TV, it is possible that some of these 'deficiencies' may be partly appeased - thus marking a perceived break with the tensions of the past (see King, 2002, p.238). As Steve Neale suggests:

> Until the recent advent of widescreen television, itself a sign of the synergy that now exists between film, television and video industries, the proportions of the television screen and all widescreen formats were significantly different - which of course was one of the reasons for Hollywood's adoption of widescreen formats in the first place

(Neale, 1998, p.130).

Yet this presents us with a trajectory in which the more recent technological convergence between film and television marks a *deviation* from the dichotomies of the past, particularly, it would seem, when it comes to the 'definitive' period of the 1950s. As suggested in the Introduction, this is part of a much wider argument which plays up conflict between individualised industries and media forms, a structure which obscures the possibility of their interaction (Hilmes, 1990). Part of the problem with previous accounts is that issues of technological difference between film and television are abstracted, isolated, and wrenched from the wider consideration of their aesthetic interplay in these early years. In terms of the cinema programme, although the coverage of star interviews, premieres or 'behind-the-scenes' footage were all exploring ways of making film 'televisual', it was around the quite literal interaction between film and television - the use of film clips - that debates around aesthetic and technological 'compatibility' were played out most clearly.

'You won't forget to say it's a *colour* film....?': Back to black and white

Without wanting to endorse an economic determinist argument, it is generally agreed that it was predominantly economic imperatives which prompted

Hollywood to increasingly adopt colour and widescreen technologies in the 1950s as a bid to stem the decline in audience attendance (Maltby, 1995, p.156). Although Cinemascope was not 'a piece of technology that had sat on the shelf for 25 years, waiting for a suitable set of economic circumstances' (Ibid), but rather also relied on post-war technological developments in areas such as sound, film stock and exhibition materials, widescreen innovations were nevertheless developed in their basic form decades before (see Belton, 1992). Given that CinemaScope did not emerge until 1953, just after *Current Release* went off air, discussions surrounding the essential 'incompatibility' between film and television initially focused on the increasing use of colour film. In the early 1950s, Hollywood's production of colour films jumped from 20% to 50% of total US output (Balio, 1985, p.425), while the British industry's take up of colour was slower and more sporadic. In contemplating the appearance of such films on the cinema programme, *Kine Weekly* emphasised the film industry's concerns over their products facing the 'disability of being "boxed up" in black and white... and viewed by the light of the flickering fire'.[1] These concerns simultaneously emerged from the protracted debates concerning the sale of full-length feature films to television. As John Davis, Managing Director of the Rank Organisation explained: 'If televised, feature films will lose much of their quality and definition and in fact give the impression that feature films are not good' (quoted in Macnab, 1993, p.201). It was clearly perceived that, when screened on television, films simply promoted cinemagoing more generally. This view was again in evidence in *Kine Weekly* when the BBC's Philip Dorte attempted to convince the film industry of the merits of broadcasting feature films: '[We are] now showing some very old feature films on television. This will remind the public of what the kinema has to offer'.[2] Interesting in this respect is the extent to which, while innovations such as colour and CinemaScope were intended to emphasise the shift to the 'event' status of cinemagoing and the specificity of the individual product, the above conception constructs filmgoing as a habitual, indiscriminate pursuit: the 'poor' appearance of a film on television - even films from decades ago - is seen to offer a 'bad advertisement' for the cinema in general, although we might note that this was at the beginning of the decade when the decline in cinema admissions was less steep (Stafford, 2001, p.99). The stakes, then, were surely far higher where the promotion of the *current* film was concerned, and the suggestion that such coverage might be counter-productive for the film industry was repeatedly raised.

At a meeting between the BBC and the film industry in May 1952, three months after *Current Release* began, the minutes record an objection from the film companies that:

> not sufficient emphasis was put on the fact that a film was in colour, and in conse-
> quence the public might miss the announcement and consider the quality of the
> extracts rather poor. [The producer] Mr Farquarson-Small took note of this, and

stated that he would do all in his power to ensure that adequate emphasis was given to these announcements.[3]

Within this context, the film industry's insistence that the presenter take due care to emphasise the status of the colour films was approached in a rather interesting and theatrical manner. In the first two editions of *Current Release* the presenter, John Fitzgerald, introduced the films from the studio 'office', complete with a young female 'secretary' named 'Susan'. According to the scripts, Susan was to interrupt his commentary - while looking over his shoulder - with such remarks as: 'You won't forget to say it's a *colour* film, will you John?' It is perhaps not only from a retrospective point of view that this strategy appears extremely self-conscious, and it emphasised the very 'newness' and fragility of these relations between cinema and television at the time. In the early 1950s it would seem that this heavy discourse of competition, so prevalent in other contexts, is the 'unspoken' in the cinema programme. It silently surrounded its rhetoric and address, providing a perspective *against* which it was keen to foreground amicable relations with the cinema. The discursive construction of colour films can be seen in this context. In fact, the film industry's concern regarding the extent to which TV viewers may consider the quality of the film clips 'rather poor' is in some sense at odds with the wider discourse of competition. If film industry marketing strategies were increasingly differentiating film from television, then audiences, while certainly new to television, were surely able to make this distinction when the clips appeared on the small screen. Nevertheless, this precisely encapsulates the contradictory position and strategies of the film industry: the desire to distinguish the cinema from television in a struggle to redefine its boundaries, yet the simultaneous impetus to use the new medium in order to promote this shift. These contradictions were intensified with the institutionalisation of widescreen technologies.

'You'd Literally Need Three Television Screens Side by Side to See It': Experiencing Widescreen Cinema on 'the Box'

Although there were a range of widescreen processes in use in the 1950s, it was Twentieth Century Fox's (TCF) Cinemascope which became most widely adopted by the Hollywood film industry, until it was eclipsed as a special event at the end of the decade by formats such as VistaVision and Todd-AO. Spryros Skouras, President of TCF, took the decision to push the adoption of CinemaScope as a way of saving his company's declining fortunes which, unlike the more short-lived 'gimmicks' of Cinerama and 3D, was perceived as the system 'best suited to regularly turning out the more expensive pictures necessary to reverse the decline in box office receipts' (Balio, 1990, p.27). Existing work on its establishment in Britain has tended to emphasise the wrangles over exhibition, such as the reluctance of the Rank Organisation (owning the Odeon and Gaumont cinema chains) and small exhibitors to make the expensive changes necessary to screen

CinemaScope films with stereophonic sound (Murphy, 1992, Macnab, 1993). Despite these problems, by the end of 1955 more than half the cinemas in Britain, with nearly three-quarters of the seating capacity, were equipped to screen widescreen films (Murphy, 1993, p.103). Much less has been said about the British film industry's adoption of widescreen at the level of production. Although colour was used for many British films in the 1950s, it was, as indicated, still usual to shoot them in black and white, as well as 'flat', a practice that continued well into the 1960s. Nevertheless, the British industry clearly experimented with CinemaScope and other widescreen technologies, early examples being *The Deep Blue Sea* (Anatole Litvak, 1955) and *Oh Rosalinda!* (Michael Powell/ Emeric Pressburger, 1955) (both shot in CinemaScope). George Perry describes *The Deep Blue Sea* (starring Vivien Leigh and Kenneth More) as an 'early casualty of widescreen' in so far as the 'melancholic drama' (set in a bedsit) was 'swamped' by the new process and unsuited to its expanding aesthetics (1974, p.45). In view of the fact that the widescreen processes were initially associated with the large-scale spectacles offered by Hollywood, this was also a recurrent and predictable response to the British use of the technology at the time. Particularly in terms of period costume features, biblical epics and musicals, Hollywood genres and budgets were deemed what Belton describes as more 'suitable for spectacularization' (1992, p.189), although this simultaneously emphasises how it is a particular *type* of picture which is used to construct the aesthetic and technological relations between cinema and television in this period. Nevertheless, using a variety of processes ranging from TCF's CinemaScope, Paramount's VistaVision, to processes such as British CameraScope, SpectaScope and SuperScope, 1956 saw several British widescreen films previewed on the cinema programme, ranging from musicals (*Stars in Your Eyes* [Maurice Elvey, 1956]), comedies (*Three Men in a Boat* [Jack Clayton, 1956]) to war films (*Battle of the River plate* [Michael Powell/ Emeric Pressburger, 1956]). While it may have been the case that British films were seen as less suited to exploitation by widescreen innovations, that is not to suggest that, with respect to the British box office, the British film was simply the poor relation of the Hollywood 'event'. As *Kine Weekly* reported in 1956, 'ironically enough some of the biggest box-office winners last year were films shot in standard ratio and without the benefit of colour'.[4]

Gomery and Allen explain how new technologies go through a process of invention (when the technology is designed), innovation (when adopted for practical use) and diffusion (when this use becomes widespread) (1985, p.114). In this respect, television played a role in the 'diffusion' of widescreen cinema, essentially from the mid-1950s. The advent of widescreen was in itself evidence of the shifts in cinemagoing and the cinema's social role, something which the cinema programme could no longer ignore. For example, *Picture Parade*'s discussion of Cinerama in 1956 was introduced by Derek Bond:

When the post-war boom at the box-office began to wane and the competition...
[from] television became more acute, the film industry made a tremendous effort in
new techniques - 3D, widescreen, VistaVision and so on. They have revolutionised
the cinema almost as much as the advent of sound....[5]

Firstly, unlike *Current Release* in 1952, it is clearly no longer possible for the
rhetoric of the cinema programme to ignore the discourse of competitive rivalry
between television and cinema. The presenter 'narrates' the dynamics of these
relations with a clarity and understanding we would perhaps expect only from
subsequent histories of the period, and the benefit of retrospection. Yet in
discussing widescreen's attempt to draw audiences back to the cinema when 'the
competition from television became more acute', *Picture Parade* presents this
strategy so casually as to undercut the grand claims of competitive rivalry. This is
reinforced by the extent to which the programme positions itself as almost an
objective observer rather than, as a *television* programme, fully implicated within
the situation described. It is from this position that, while clearly less securely than
earlier in the decade, the cinema programme continues to see the media as
companions rather than antagonists.

Following the general introduction, Bond goes on to describe Cinerama as 'the new
form of entertainment' in which the 'all-embracing screen gives a new thrilling
experience'. He then previews the film *The Seven Wonders of the World* (1956)
which was the latest release in the process (although given that Bond introduces
Cinerama as 'a new form of entertainment' it would appear that Cinerama was
introduced to Britain some time later than in the US, where it had emerged in
1952). In the preview on *Picture Parade*, viewers were shown an excerpt from *The
Seven Wonders of the World* which, like many Cinerama features, was a 'sight-
seeing excursion' and 'compendium of exotic locales', visiting places such as the
Taj Mahal, the Pyramids and the Leaning Tower of Pisa (Belton, 1992, p.91). In
fact, while it may be the case that television failed to truly offer the sense that 'the
all-embracing' Cinerama 'screen gives a new thrilling experience', there is
simultaneously a way in which this presentation was entirely in keeping with the
discourses which surrounded early television viewing. Belton has argued that the
emergence of widescreen cinema calls for a 'rethinking of traditional models of
spectatorship' (1992, p.185). The aesthetics, but particularly the marketing of
widescreen, was shaped not only by competition from television, as in order to
compete with the proliferation of other leisure activities, films had to redefine
themselves as a 'more participatory experience in which they became not so much
something people saw as something they *did*' [original emphasis] (1992, p.95).
Cinerama in particular drew upon the recreational referent of the amusement park,
the 'visceral thrills' of which have conceptual connections with early cinema
(Belton, 1992, p.188). This is also the case with Cinerama's penchant for
'actualities' and travelogues. In its rejection of character and narrative, the

emphasis in Cinerama was on the spectacle of the technology, and its ability to define cinema as 'pure spectacle, pure sensation, pure experience' (Belton, 1992, p.97). Of all the widescreen innovations, the marketing for Cinerama drew most strongly on a sense of audience involvement, and adverts promised: 'you won't be gazing at the movie screen - you'll find yourself swept right *into* the picture, surrounded by sight and sound' (Belton, 1992, p.92). It is in this respect that Belton claims that, in its bid to aesthetically and discursively define a new viewing subject for the medium, the motion picture sought a 'middle ground between the notion of passive consumption associated with at-home television viewing and that of active participation involved in outdoor recreational activity' (1992, p.187).

In fact, rather than such strategies 'clearly distinguish[ing] cinema from television' (Belton, 1992, p.192), they actually *echo* its discursive construction quite clearly. As discussed in Chapter Six in relation to the film premiere, key was not simply the emphasis on television as a live medium, but the ways in which this set discursive parameters for viewer interaction - and the use of metaphors of transport and travel were crucial here. Belton's emphasis on widescreen engendering of a 'co-presence' between spectator and screen, breaking down the traditional 'segregation of spaces' (Belton, 1992, p.192), clearly mirrors the tropes used to describe early television, in effect drawing on a discourse of *liveness*. In terms of the cinema, this in itself was not of course intrinsically new. In his discussion of historically 'repositioning cinema and the viewing subject within a field of televisual expectation', William Uricchio explains how cinema historians have tended to 'flatten the discourse of liveness' in the development of the cinema (1998, p.19), and the extent to which it structured the expectations within which the medium emerged. (As Uricchio notes, the 'romanticised retelling of the "Lumiere effect"' - an impression of reality so strong that early film viewers allegedly sought cover from images of an oncoming train, might be seen in this context). The cinema could not ultimately deliver 'simultaneity over distance', but these discourses of liveness are clearly re-invoked in the promise of widescreen technologies decades later. That is not to imply that the experience of early 'tele-viewing' and widescreen spectatorship were in actual fact 'really' the same. Rather, it was that, in jockeying for a place in the ongoing redefinition of screen entertainment in the post-war period and its shifting relations with public and private spheres (see Spigel, 1992), the discursive construction of cinema and television in some ways converged, rather than diverged. It from this perspective, particularly given Cinerama's interest in the travelogue, that *Picture Parade*'s preview of *The Seven Wonders of the World* is in many ways entirely appropriate to the textuality and experience of early television.

On occasion, however, there was clearly no way to avoid these technological 'divisions' between the media, as *Picture Parade*'s preview of *The Windjammer* (1956) makes clear. A semi-documentary feature about a Norwegian sailing vessel, this was a Cinerama-type film shot in the latest widescreen process, Cinemiracle. (The less than exciting subject matter here indicates the primary importance given to technological process and form). *Picture Parade*'s Peter Haigh opened the programme with the preview:

> To start off this week, I'd like to tell you a little about a film that we can't show you an extract from, because you'd literally need three television screens side by side in order to see it... Just to give you some idea of how wide the picture is, here it is compared with the normal screen [photo caption] that's the one at the top - and CinemaScope [photo caption] that's the one in the middle. It looks pretty impressive in the cinema - the screen goes right across from one side to the other. And in order to get such a wide picture they have to use a special camera, with three lenses working at once, photographing it in sections, so to speak.[6]

Haigh's description pivots entirely on scope and size, the width of the screen. This is differentiated not only from television, but from the conventional cinema screen, as well as existing widescreen processes such as CinemaScope. As Belton explains, the three lenses of the Cinerama camera were crucial in engendering the sense of audience participation. The lenses encompassed an angle nearly approximating that of human vision, reinforcing the encompassing illusion of three-dimensionality (1992, p.92). This, however, also created problems in exhibition, both in terms of the appearance of seams (where the three images were joined), and in allocating space in the cinema to accommodate the three projection points. It is thus not difficult to see why the BBC needed to find an alternative method for previewing a film which they *literally* couldn't screen. After the introduction to the three-lens camera, the viewer is shown a photograph of the 'mysterious' technology ('It looks pretty complicated to me'), and this is followed by stills from the film itself: '[Photo of ship] There she is - and very graceful she looks too. [Photo] That's one of the rougher moments when they hit a howling gale in the Atlantic'. Forced to rely upon the photographic image, television is in fact returned to the pre-cinematic - apparently so different are the media that sharing material is impossible. As the preview draws to a close the viewer has to rely upon the presenter's *recollection* of the event: 'The film is in colour and when I saw it at the London premiere... it certainly looked impressive...' A similar example can still be seen later in the decade with *Ben Hur* (Sam Zimbalist, 1959) when the (new) presenter, Robert Robinson, explained that: 'we were going to show you something from the film, but we can't'. Waving the reel of film in front of the camera he elaborated that it was simply 'too big. We use 35mm and this is filmed in what is called Camera 65...' Instead, Robinson spoke to the film's musical composer, Miklos Rozsa.

While in each case specific films are involved, these examples also explore particular widescreen processes. Such cases were vastly outnumbered by the regular previewing of CinemaScope films. Many companies supplied monochrome prints of CinemaScope films which offered a better definition on the television screen, demonstrating the extent to which, despite their complaints, the film industry were willing to invest carefully in television's promotional opportunities. Yet just as with the issue of colour films, the film companies insisted that the programmes stress when a film was shot in CinemaScope (or other widescreen formats) so as not to allow television's 'deficiencies' to obscure this fact. If such announcements did not occur, the seriousness of the matter would be brought to the programme makers' attention. Associated British Pathé, the distribution arm of the Associated British Picture Corporation, found extreme cause for complaint when *Picture Parade* previewed the British comedy, *Let's Be Happy* (Henry Levin, 1957). The Publicity Manager wrote to the Producer of the programme to explain his case:

> When it was decided to prepare, on your behalf, black and white films in order that you could have the most satisfactory pictures, I spoke to your secretary and asked whether you wished us to supply these in unsqueezed form. I was immediately advised that there was no necessity for this because the new apparatus you have installed overcomes the difficulty of the cinemascope ratio. Despite this assurance, however, I understand that the prints were shown with the top and bottom of the screen empty. This could have been avoided had we not received the assurance mentioned above. Finally, no mention was made to the effect that the picture is both Technicolor and CinemaScope, two refinements which we consider still have an effect upon the box office.[7]

The aspect ratio of CinemaScope, of course, posed problems for television broadcast. If the full width of the widescreen image were shown, then the top and bottom of the screen would be masked. If the top and bottom of the screen were used, then the sides of the image would be cropped by up to 50% or more (Belton, 1992, p.216). Rival programme *Film Fanfare* was also faced with this situation, and it was at one point very frank about these difficulties. Presenter John Fitzgerald saw it as his duty to explain the technological conflict to the audience:

> For the benefit of new viewers I should explain that to present the... shape of the CinemaScope screen on television we have to have a lot of black masking at the top and bottom of our screens and on the other hand to fill our screen means clipping a lot off the sides of the movie picture. There is a lot of argument as to how this should be done. Perhaps the arguments will be settled one day when we get elastic television tubes. Meanwhile, we compromise by clipping a little bit off the sides of the picture and having a little black masking at the top and bottom...[8]

The presenter appears to acknowledge the heated debates surrounding the film industry's concerns over the broadcasting of CinemaScope, while again still refusing to engage directly with the proposition that cinema and television may occupy an antagonistic relationship. He presents this as a temporary compromise awaiting technological advancement, while simultaneously a situation ('when we get elastic television tubes'), that is virtually *impossible* to overcome.

As Belton explains, with regard to the early 1960s, the major American TV networks did in fact quickly '"resolve" the essential incompatibility between the widescreen and television formats by deciding to pan and scan (that is, to crop) films rather than to present them in what was subsequently referred to as the "letterbox" format' (Belton, 1992, p.216). Yet British television companies also appear to have been venturing into technological solutions well before this time. As the letter from Associated British Pathé to the BBC makes clear, there had been talk in 1956-57 about developments enabling the televising of CinemaScope films. In January, 1956, *Today's Cinema* reported that CinemaScope films could now be broadcast without the masking on the TV screen.[9] However, given the film industry's reluctance to sell films to television (and the fact that, unlike the films that were screened, CinemaScope features would be recent products), it seems very unlikely that many were broadcast in full at this time. Just before this announcement was made, the CinemaScope film *The Black Widow* (Nunnally Johnson, 1954) was in fact screened on ITV. Nevertheless, this murder mystery (starring Ginger Rogers and Van Heflin) was evidently perceived to be something of a special occasion, screened as it was during the Christmas period at 3pm in the afternoon. Pioneered by ITV television company Associated Rediffusion, it is not entirely clear what system was in use here, but it was clear that strategies were being considered fairly swiftly. While television was focused on the long-term possibilities of interaction between cinema and television, the film companies seem to remain fixated on the construction of a different exchange value for the filmic product. This was of course paradoxically accentuated by the very problems about which they complained: television's inability to present the CinemaScope image *'proper'*. While it clearly heightened awareness of the film's release, television simultaneously displayed the distinctions between the media which the film industry were at pains to accentuate. The cinema programme operated here through a kind of 'double language': it promoted the new processes by insisting on their advantageous effect on the cinemagoing experience while through this very act - whether with respect to the appearance of the masked, monochrome image or the dutiful verbal emphasis on the 'spectacle' television could not display - differentiation was secured.

The precarious nature of these relations was emphasised in 1957-58. In 1957, Twentieth Century Fox and Warner Bros, as well as the Rank Organisation (the three companies which, aside from MGM had contributed most to the British cinema programme), announced that they would no longer provide clips for the BBC's *Picture Parade*. Warner Bros and Fox claimed that, in view of television's continued threat to cinema attendance in the US, this was taken on 'orders from America'.[10] Yet Fox and Rank in particular claimed that their decision was primarily due to the BBC's inefficiency in screening CinemaScope films. Interesting here, however, is how the emphasis on CinemaScope as a reason for withdrawing from television seems to prefigure the dichotomies which permeate subsequent critical histories of the period. The fetishisation of the technological 'incompatibilities' between the media are used as a 'smokescreen' for what are essentially a much broader set of prejudices concerning the aesthetic relations between cinema and television.

A key question is whether the cinema programme demands a reconceptualisation of the established historical narrative concerning the aesthetic and technological identities of film and television in the 1950s, and the essentialist claims on which they are based. Although emphasising the cinema's increasing dependency on television for its cultural visibility, it is clear that the cinema programme often supported the film industry's marketing strategies at this time - their bid to differentiate cinema and television. Yet what is also clear is that television seemed entirely happy to occupy this position: although the relations between the media, in the earlier part of the decade at least, were open to question and not perceived as fixed on a definitive future course, television had nothing to fear from the cinema. After all, in its very production, articulation and audience address, the cinema programme emphasised that, despite ventures into such technological appeals, the audience was watching the small monochrome screen at home. This does undercut the claims of the conventional historical trajectory in this respect, precisely because it makes it difficult to see the media as individualised institutions, structures and textual forms.

It is, of course, relatively difficult to gauge the effects of their relationship in this respect, although it is worth noting that a survey conducted by *Kine Weekly* in 1956 reported that the most frequent factor cited as prompting people to see a film was 'film extracts on TV programmes'.[11] Equally, 88% of patrons questioned indicated that widescreen innovations had increased their enjoyment of film - the two areas clearly not being mutually exclusive. We also know comparatively little about the *reception* of new cinema technologies - ranging from sound, widescreen to more contemporary examples such as CGI - given that technological studies tend to concentrate on industry and text, rather than reception and audience. Certainly, from a historical point of view, this is not an easy area to research, yet there are ways of considering the 'diffusion' of technologies in wider cultural and media contexts. In terms of colour and CinemaScope, 1950s television tells us just as much about the circulation and consumption of cinema as the films themselves, or the evidence of box office receipts. Equally, historical traces indicate that there are further unanswered questions and avenues to pursue. What is the context, for example, for *Picturegoer*'s report in 1953 that a Hollywood TV producer was shooting a series of

3D films for viewing in the home through 3D glasses?[12] Finally, perhaps it questions the conception that colour and widescreen simply emphasised 'the "specialness" of cinemagoing... [in] stark contrast to the banality of staying at home and watching the "box"' (Macnab, 1993, p. 211). Was television in the 1950s - when for many its experience was still new - ever simply 'banal'? Perhaps not. As one viewer put it in a cinemagoing survey in *The Advertiser* in 1957: 'My husband has recently bought a 17" television set. With that screen size, you can keep your widescreen films'.[13] Widescreen TV here we come?

'The Infant Medium with the Adult Manner': The 'X' Film on Television

As discussed as the start of the chapter, technology was not the only means by which the film industry aimed to differentiate their product from television in the 1950s: film content is also crucial here.

In 1962 the *Sunday Independent* complained about the BBC's cinema programme, *Picture Parade* (1956-62), and its repeated promotion of films unsuitable for family viewing:

> Picture Parade is still as dirty as ever. The persistence with which this feature selects all the suggestive bedroom scenes is remarkable and consistent. The BBC would not dare to screen a programme of this kind if Sir John Reith were at the head of its affairs. There has been a sad deterioration since he left.[14]

Shaped by the prevailing climate of the 'permissive society', *Picture Parade*'s interest in such films is seen as symptomatic of a more general decline in the BBC's moral standards. The use of the term 'still as dirty as ever' suggests that this is not an isolated complaint, and it assumes a shared knowledge between critic and reader that this is 'typical' of *Picture Parade* and indeed the BBC. According to the *Sunday Independent*, television is actively involved in - even specifically *associated* with - the cultural circulation of 'risqué' films. This begs the question as to how the relations between British film culture and television had reached the point when such examples were 'typical' and a subject of moral concern, particularly when films were meant to be offering the very antithesis to their small screen rival.

An indication of this complexity is given in James C. Robertson's observation that media censorship relaxes as the importance of that medium declines, a factor which contributed to the British Board of Film Censors' (BBFC) increasing liberalism in the 1960s when the cinema was moving away from its status as the most popular mass medium (1993, p.159). It follows, then, that just as the earliest years of cinema were free from stringent censorship, so the early years of television were characterised by a *less* cohesive and institutionalised form of regulation. That is, strategies and rules were still being formed. The 1950s were a time when, although beginning to decline, the cinema was still a popular pursuit, and television was just rising to its status as a mass medium. As a result, it was the site of some interesting contradictions where the regulation of film content was concerned.

This in itself offers a challenging perspective as the BBC is generally seen as more conservative, didactic and paternalistic than its commercial counterpart, but it was the Corporation that was at the forefront of these relations with film. That is not to suggest that the BBC's cinema programmes deliberately pursued coverage of X-rated films. The story of their promotion is simultaneously the story of how British television developed policies for *regulating* film content, strategies clearly shaped by institutional policies of public service. To a certain extent, it is problematic to isolate examples of the X-rated film as they are the product of a subjective classification which encompasses different types of films. However, when exploring these relations, it is necessary to focus on particular examples, and it is through the X-rated film that the policies of both the BBC and the BBFC are most visible. These policies are traced by considering how X-rated films first appeared on the cinema programme, and the controversies and regulatory policies which followed. This is followed by *Picture Parade*'s role in previewing social problem and Continental films, before a consideration of how television's relations with 'adult' films intensified by the early 1960s as the X certificate moved into the mainstream of British film production.

'A tawdry lesson in easy living': *La Ronde* on the BBC
In terms of general programme content, the BBC developed their own forms of regulation where issues of 'taste', 'decency' and controversial material were concerned. From the early days of radio the Corporation monitored programming very closely and as Lord Reith stressed in 1923, 'with approximately two million people, including children, listening, it is obvious that all matter must be entirely above reproach' (Shaw, 1999, p.23). Practicing a form of self-censorship, producers worked within strict internal guidelines, a system also adopted by television. In fact, the Controller of Television Programmes, Cecil Mc Givern, emphasised that in handling sensitive or controversial material, 'we in television must ...be more careful than "sound" producers' and cited the medium's visual capacity in support of this (Briggs, 1975, p.212). When relating this perspective to the feature film, it is clear that their regulation on television posed problems in ways that radio had not. Radio, of course, lacked a visual dimension, and while film soundtracks are equally subject to censorship, the image is given far more attention. The trend in 'adult' films often traded upon their visual appeal, and unlike television, radio could not be accused of showing sexually explicit, violent or 'harmful' images. Furthermore, given that they were partly a response to the new medium, the representations in question primarily emerged in the post-television age. The BBFC instituted the 'X' certificate in 1951 to embrace all features deemed unsuitable for under 16s, and it was originally intended to encourage the

production of 'adult' orientated pictures which previously would not have reached the cinema screen (Dewe Mathews, 1994, p.25). The secretary of the BBFC, Arthur Watkins, expressed his belief in the protective, as well as the artistic value of the X certificate, and foregrounded its 'cultural value' in opening 'up all kinds of possibility [sic] to imaginative producers'.[15] This was not necessarily the reality of the situation as distributors and exhibitors still favoured the concept of 'family entertainment', and were less willing to release, screen or finance X-rated films, although as explained below, exhibition strategies also began to change (Aldgate, 1995, p.14).

Many of the first X certificate films were foreign-language features and they enjoyed an expanding circulation in Britain as the decade progressed. In 1952 this was still in its early stages - the year that the BBC were first confronted with the issue of previewing films unsuitable for the family audience; films which, when screened at the cinema, carried an X certificate. *Current Release* was essentially a preview of the latest popular American and British product and did not often feature foreign-language films. Nevertheless, in the second edition of the programme viewers were introduced to the X-rated French success, *La Ronde* (Max Ophuls, 1950). *La Ronde* had been running in London at a specialised cinema for one year and it was also screened in some provincial cities. Its British exhibition was now set to range more widely which perhaps explains its appearance on *Current Release*, a full year after its British premiere. This now classic film told a tale of the circuitous love affairs and infidelities of occupants of 1900 Vienna (hence the title *La Ronde*, French for 'roundabout'). The film's depiction of the 'roundabout of love', however, earned it an X certificate, and it received a controversial reception in both Britain and America. In the US it was banned in many cities and tried in the courts for obscenity (where the prosecution enquired, 'do you realise that this picture will be seen by a large group of people in a darkened theatre?') (Dewe Mathews, 1994, p.127), although the US Supreme Court later found in favour of *La Ronde* in light of that year's decision regarding *The Miracle* (Roberto Rossellini, 1948) - that films are entitled to the constitutional guarantees of free speech (Mc Donagh,1996, p.137). As tame as it may seem today, it thus arrived in Britain with an already controversial status, and many critics were shocked at the prospect of its wider exhibition. An article in *People* expressed particular alarm and ran the headline 'Would you let your daughter see this?': 'when all the brilliant art of its acting and production are taken away, this is nothing more than a bawdy sex film that glorifies promiscuity and mocks at marriage'.[16] After consulting the opinion of an 'expert' psychologist, the article claimed that it gave young people a 'tawdry lesson in easy living' and urged local councils to ban the film.

It may seem surprising, therefore, that in the second edition of *Current Release* the presenter, John Fitzgerald, cheerfully introduced this now notorious product. As though to justify this move he emphasised its accolades and awards, since as the comments in *People* make clear, *La Ronde* had also acquired a reputation of artistic merit. Foreign-language films, of course, have been associated with something of a dual identity since the early days of cinema. They connote artistic prestige, yet simultaneously, are associated with the 'risqué' and the sexually explicit. Given that it represented the main theme of the film, *Current Release* showed scenes offering glimpses of the love affairs, yet Fitzgerald's commentary made no reference to its controversial status and (following the excerpts), foregrounded its 'adult' associations from a different perspective: 'Very grown-up - very sophisticated - with a lot of shrewd reflections on the social attitudes of that Old Vienna of 1900. And isn't that tune haunting....?' The Head of BBC Television Film later criticised the preview on several counts, although this had little to do with the subject matter of the film. Rather, it was partly because it was not on general release and that 'the BBC has no difficulty in getting foreign-language pictures... we could get *La Ronde* for transmission in toto'.[17] As discussed, this suggests that at a time when full-length pictures on television were in short supply, part of the appeal of the cinema programme was its inclusion of the 'forbidden' product of the feature film. Particularly interesting here, however, is the BBC's belief that *La Ronde* - the film that could be seen to offer 'a tawdry lesson in easy living' - would be an entirely *acceptable* proposition for a feature-length screening.

It was not in fact the BBC but the film industry which noted the potential problems raised by the preview of the film. The *Daily Film Renter* remarked that it was the first X film ever to appear on television[18] and *Kine Weekly* complained that nothing was shown to indicate its censorship rating, and thus that children under 16 could not be admitted to a theatrical screening of the film.[19] After noting that *La Ronde* was given the 'choice position in the... programme and the longest screentime' the critic suggested that 'it would be a good idea if the BBFC certificates were shown on the final title'. This perhaps suggests less a fear for the moral health of the juvenile audience than a concern that television may be undermining established censorship regulations which, if an influx of younger patrons attempt to gain access to an X-rated film, may pose problems for the exhibitor.

'Truly Shocking': The *Rashomon* Controversy

These fairly muted, although significant, concerns paved the way for a much more heated controversy two months later. The BBC appeared to remain largely unaware of any potential problems until the sixth edition of *Current Release* when the Japanese success *Rashomon* (Akira Kurosawa, 1951) was previewed. This now renowned feature is set in medieval Japan and explores how four people have different versions of a violent incident when a bandit attacks a nobleman in the forest. The story is then told four times from each point of view. *Rashomon* also included a highly controversial scene depicting the rape of a young Japanese girl and although passed uncut, the BBFC awarded it an X certificate (Dewe Mathews, 1994, p.127). *Rashomon* introduced Japanese films to the world market and many British critics were unsure quite what to make of it, expressing a mixture of uncertainty, bewilderment and admiration. However, although unaccustomed to Japanese cinema, many agreed that it was indeed a 'masterpiece',[20] and when included in *Current Release*, *Rashomon* - like *La Ronde* - was the focal point of the programme. A pattern emerges here as the presenter introduced *Rashomon* by stressing its artistic prestige and accolades (including an Academy Award for best foreign film and the top international award at the Venice Film Festival). Again, this was partly invoked to justify the BBC's preview of the film while simultaneously, also reflecting on their selective criteria where the previewing of possibly sensitive material was concerned. Due to their perceived lack of familiarity with Japanese cinema and inadequate television subtitles, Fitzgerald talked the audience through the chosen scenes. He then introduced someone who was constructed as an 'expert' speaker to discuss the film:

> Now this film made a very deep impression on most of the critics, but some of them added a footnote to the effect that they regretted a lack of terms of reference to the Japanese mind and language, to their culture... It's a very good point, and to meet it we have in the studio with us ... a man who knows Japan. He's written several books about it. He also happens to be the head of the BBC's Far Eastern Service.[21]

This was Mr. John Morris who, having lived in the country for some time, claimed to be familiar with Japanese cinema. He went on to discuss a range of issues including the film's literary background, acting in Japanese cinema, and compared *Rashomon* to other Japanese films of its time. With the help of the BBC's Far Eastern Chief, *Rashomon* is constructed through a reflective, serious and almost intellectual gaze - a very selective perspective in which no mention is made of the film's controversial content. However, this presentation did little to disguise the issue, and one viewer was so incensed that she wrote to the press and the BBFC to complain. The BBFC then forwarded her letter to the BBC:

> I wish to make the strongest possible protest at the showing of... an X film in... *Current Release*... The film was Rashomon and I should imagine the very part for which the film was given an X was shown. I have two boys of 12 and 15 - most impressionable ages - and *Current Release* was put on at 8:15, too early for... [them] to be in bed. My boys were waiting to see a comical Magician, Tommy Cooper, which was foolishly put on after *Current Release* and I really think it is a scandalous thing that an X film should be shown over the television where simply hundreds of children must have viewed it... the whole purpose of the X [certificate] is defeated. The effect of watching the rape of this Japanese girl on these young and sensitive minds must have been truly shocking.[22]

Individual responses are seldom on their own sufficient evidence with which to explore the reception of a film, and the letter could just as easily be seen as an interesting anecdote - an insight into the sexual and social morals of the time. Yet this letter, and the ensuing dialogue between the BBC and the BBFC, *prompted* the Corporation to address the issue of regulating film on television, and to consider its institutional and cultural implications. The viewer's letter raises several key points about this issue. Although she seems concerned that such material was shown *at all*, she particularly emphasises when it was screened and thus where it appeared within television's daily flow. Her children only saw the scenes while waiting for an entertainment programme and *Current Release* was screened earlier on this occasion, 8:15 rather than 9:30pm. Although to this particular viewer the policy already seems common sense, the letter indicates that the concept of the 'watershed' - that all programmes before 9 o'clock be suitable for family viewing - was not yet established, and did not in fact emerge until the end of the decade. She evidently perceives that such images are more accessible on television where (unlike the cinema), the audience is not regulated in any way. Perhaps most crucially it attests to the confusion surrounding the institutional responsibility for film content once moved from its primary site of exhibition. The viewer writes not to the BBC but to the press and the BBFC. She perceives the BBFC to be responsible for the regulation of film and as a result, she directs her complaint to them.

While the BBC may have developed means of regulating their own material, films represented something 'other'. They were acquired rather than produced in-house, and were classified and regulated by a body outside of the BBC. Now, however, films could be shown through an alternative distribution outlet, whether in whole or part. Were the BBC to abide by the BBFC's classifications, and if so, how? What responsibilities did they have, and how should these be defined? At this point they were caught without an official policy of any sort and this was something of an embarrassment for the BBC, especially when television was attempting to establish cultural respectability. In 1952 the X certificate was still relatively new, and its implications were monitored closely by the BBFC. In forwarding the complaint to the BBC, they were clearly concerned about the *Rashomon* incident and siding with the viewer, Arthur Watkins bluntly informed the BBC that there was 'some point in the writer's letter which [they] might care to consider'.[23] The producer of *Current Release* was defensive and refused to recognise the implications of the situation. He maintained that the extracts were 'carefully edited' so as not to cause offense and in emphasising that evening programmes were designed exclusively for adults, he placed rather more onus on the parent's responsibility than he did the BBC's. This was again stressed in the BBC's official reply to the BBFC in which they maintained that *Current Release* had screened what they considered to be an 'A' extract from an X film, 'aimed exclusively at an adult evening audience'.[24] However, the Programme Controller elsewhere noted that 'the implications [of the situation] are ... important' and suggested that the producer was evading the issue 'since, through an extract, we showed the "big scene"'.[25] Given perceptions of the BBC's paternalistic and conservative moral values, the image of the presenter introducing these scenes is quite startling, and attests to the fact that the Corporation had not really given the regulation of film on television any considerable thought.

While the BBFC classify films in whole rather than in part, X films could be advertised by an A or U trailer, as the BBC rightly suggest. However, the key question is how the BBC, an entirely unofficial body where film censorship was concerned, could *decide* which extracts deserved which rating. An internal memo explained how although they were aware of the 'responsibility of the X certificate', the Corporation did not recognise the BBFC's classifications in any way. In fact, it was precisely the BBC's perception of their *own* responsibilities which deterred them from acknowledging the official categories. It was explained that 'the policy about showing an A or U extract from an X film has not in the past been clearly defined ...[as] there has been anxiety lest categorising a film as 'X' might create an unhealthy interest'.[26] The BBC did not wish to be perceived as 'enticing' the public to view such material and this reflects on the extent to which the X certificate had already acquired dubious connotations. In 1952 *Picturegoer* magazine expressed a similar concern that the X certificate might be used as an exploitation strategy, encouraging people to queue for films in 'the hope of seeing at least as much as the butler saw...'[27] Despite this controversy, *Current Release* continued to occasionally preview X-rated films, something highlighted by the *Manchester Evening News* when it reported that 'councilors are crossing swords over the televising of "X" certificates in *Current Release*'.[28] The year 1952 was still early in the development of television and the amount of light entertainment was limited. A programme which covered the popular interest of the cinema attracted what (at the time) were large numbers of viewers. As a result, the previewing of these films was widely recognised and discussed.

However, after the *Rashomon* controversy the BBC informed the BBFC that they were 'anxious to clear up the situation' and established provisional guidelines. These were: never to show an X extract from an X-rated film, never to show an A extract from an X film without verbal warning and, finally, to ensure that the programme does not precede a feature with 'family' appeal.[29] The concept of the verbal warning is, of course, still a practice used today, yet the question remains as to how the BBC could categorise the excerpts, and with what authority. The Corporation appear to have kept their word for as a letter to *Picturegoer* enthused, 'thank goodness TV's *Current Release* warns us in time to keep away from X films'.[30] Here, rather than undermining the authority of the BBFC, the BBC is constructed as a guardian of public taste.

The BBC's reply to the BBFC emphasised how their use of the X film would be 'rare', and their selection criteria is revealing in suggesting how the regulation of film was put into practice, and its relations with public service. Tom Dewe

Mathews explains how as the 1950s progressed, the BBFC's own criteria underwent a change. A pertinent influence here was the emergence of an academic discourse on film and its interest in the director as a critical force (as associated with the writings of the *Cahiers Du Cinema* critics by the mid-1950s, although auteurism was to become more important in the following decade). According to Mathews, an emphasis on the 'auteur' tied in with the BBFC's own beliefs that one 'should give full weight to the cardinal factor in the censorship of films, the "intention" of the director' (1994, p.126). Mathews explains how the BBFC moved toward an artistic rationale for film censorship: Watkins' 'prescient belief in the auteur theory created a loophole for such sequences as the rape scenes in Kurosawa's *Rashomon* to squeeze unscathed through his office; as did the ... circular sexual couplings of Max Ophuls' *La Ronde*' (Ibid, p.127). Not coincidentally, these were of course the first two X films used in *Current Release*, and while their directors were not a particular point of emphasis, they were partly included for their artistic merit. If the BBC were to show an X picture very 'rarely' and even then, 'only because of the quality of the film',[31] they too were operating an 'artistic rationale' when it came to film regulation. It also indicates how the inclusion of the X-rated film was understood to represent a 'risk' for the BBC (in terms of attracting possible criticism), and that they were only willing to take this chance if a certain prestige was gained in return. That said, the BBC's involvement with such films would be by no means as 'rare' as they were apt to claim.

This indicates how policies regarding the regulation of film on television emerged not in relation to the full-length feature, but with regard to the preview, and the cinema programme. This was largely because, at least until the period 1957-58 when feature films entered the schedules at peak time on a Saturday night, the cinema programme was the main experience of film on British television. As a result, the use of the term 'seeing a film on television' lacked the precise meaning that it holds today. (As a critic in *Today's Cinema* noted, 'patrons insisted that they had "seen a film on television" when, in actual fact, all they had seen was an extract').[32] With respect to the films discussed so far, it is not so much what is actually *seen* in the excerpts (although the complaint about *Rashomon* was specifically linked to its content). Rather, it is more television's *association* with the promotion of such films that seems to question existing histories, and these relations intensified as the decade progressed. The BBC may have begun to establish certain policies for dealing with such material, but this did not prevent television playing a role in the cultural circulation of the X-rated film. By 1956-58 and the emergence of *Picture Parade* and *Film Fanfare*, we see a certain shift in these relations, and in the BBC's criteria for the regulation of film. This works in dialogue with changing policies in film censorship, and shifting trends in film production and exhibition.

Although (due to available evidence) discussion here largely centres on the BBC series, it is worth noting that the emergence of a second channel foregrounded the wider issue of censoring or regulating television material (see Sendall, 1982, p.327). Imported series such as *Dragnet* (ITV, 1952-59) and *Gunsmoke* (ITV, 1955-75) raised more serious concerns about the regulation of television programmes (in this case violence), particularly when it came to their 'effects' on the young. It appeared to many that there was an absence of official regulation and the complaints were raised in the House of Commons. With regard to general programmes and films it is clear that (anxious to guard against outside intervention), the BBC began to watch these issues more closely. They considered categorising programmes with an A or X certificate, as though 'they had been viewed by a censorship body observing the same kind of standards as the BBFC'.[33] With regard to feature-length films (or rather those considered to be an A certificate), it was suggested that they place an indication in the *Radio Times* which read (in capital letters) "'MORE SUITABLE FOR SHOWING TO ADULT AUDIENCES'". The BBC understood that 'this is now used by the BBFC for A films: it is not so specific as saying that a film is unsuitable, and it puts the onus on the parents to decide'. This highlights the complexities of the BBC's position. They wanted to be seen to exercise an official authority comparable to that of the BBFC. Simultaneously, however, they wished to remain an autonomous body governed by internal rule, and the fears concerning outside intervention were clear.

It is clear from reports that there were occasions in the late 1950s when X-rated films were broadcast on television in their entirety. *Today's Cinema* noted the 'amazing situation regarding film censorship: What is [its]... future if... X pictures can be shown on television to anyone, whether over 16 or not?'[34] An exhibitor later complained that 'the TV channels are allowed to show what films they like, murder, sex, brutality, without any interference from Watch Committees'.[35] So while on the cinema programme you could catch a preview or discussion of the X-rated film, they could also occasionally be viewed in full.

By mid-decade there was much discussion of the X certificate by the British film industry which felt that, rather than offering the artistic freedom envisaged by Arthur Watkins, it was clearly hindering British filmmakers. At a meeting of the British Film Producer's Association (BFPA) in 1955 it was argued that the X certificate was no longer serving the purpose for which it was intended. Rather, it was claimed, the X certificate was being used as an exploitation strategy and facilitating the wider distribution of Continental films in Britain while the attempts made by British filmmakers to produce adult films had often failed, partly because the major circuits refused to book them (Aldgate, 1995, p.18). In comparison with the Odeon, Gaumont and ABC cinemas, the BBC were less reluctant to include such films in *their* programmes, and supporting the claim that the X certificate had not aided the success of the British film, those previewed on television now tended to be American (and as I shall discuss, Continental). Where the 'risk' of the X-rated film was concerned, the BBC's *Picture Parade* tended to favour the current trend in social problem films, and in terms of their regulatory criteria, it was evidently these that were regarded as being of 'exceptional merit'. While British cinema of the late 1950s equally exhibited an interest in contemporary social issues, the censor at this time, John Nichols, is said to have shown a particular sympathy toward the British social problem film (Dewe Mathews, 1994, p.138). They were evidently perceived as less 'daring' by the BBFC, and were less likely to carry an X certificate.

'Difficult and demanding'?: Television and the social problem film
When it came to Hollywood, the social problem film was evidence of filmmakers increasingly rebelling against the constraints of the Production Code. By mid-decade Hollywood was producing films which dealt with previously forbidden themes, or which explored sensitive issues in a more explicit way. These included, for example, different forms of drug addiction (*The Man with the Golden Arm* [Otto Preminger, 1955], *Bigger than Life* [Nicholas Ray, 1956]), homosexuality (*Tea and Sympathy* [Vincente Minelli, 1956]), abortion (*Bachelor Party* [Delbert Mann, 1957]), alcoholism (*I'll Cry Tomorrow* [Daniel Mann, 1955]) and unconsummated marriage (*Baby Doll* [Elia Kazan, 1956]) (Doherty, 1988, p.33). With competition from television and declining audience attendance, the studios and exhibitors pressurised the MPAA to relax its moral standards and in December 1956, the Code was officially revised to allow references to drug addiction, abortion, prostitution and miscegenation (Klinger, 1994, p.39).

In her excellent historical and contextual study of the films of Douglas Sirk, Klinger discusses the 'local genre' of the 'adult melodrama' and emphasises how the trend toward 'daring', 'shocking' and adult themes pervaded melodrama more widely at this time. As Klinger explains, the Hollywood industry defined these films through a double language which foregrounded their social significance as justification for a 'titillating indulgence' in spectacles of alcoholism, adultery, homosexuality and rape. Their promotional campaigns called attention to their challenge to censorship and the studios constructed themselves 'as crusaders for the freedom of speech... and the right of their audiences to experience mature, realistic content' (1994, pp.40-41). While it may seem that such sensationalism emphasised the exploitation value of the X certificate - associations which the BBC were clearly keen to avoid - it is not difficult to see why the Corporation approved of films which dealt frankly and openly with contemporary social issues. To a certain extent, they could be perceived as educational, as well as entertaining. Films were adapted from prestigious plays or novel with 'celebrated adult profiles' and *Baby Doll* was one of several Tennessee Williams plays to appear on the screen (Klinger, 1994, p.40). In fact, it may well have been that viewers first encountered one of the most controversial films of the decade on television, as the showpiece in an edition of *Picture Parade*. Centering on a character with a beguiling childlike sexuality, the title role of *Baby Doll* was played by Carroll Baker who, after two years of marriage to Archie Leigh (Karl Malden), refuses to consummate their relationship. Baby Doll is then seduced by her husband's revenge-seeking rival (Eli Wallach). In fact, *Picture Parade's* extract from the film depicted Baby Doll flirting with her suitor as she swayed across the veranda and claimed in her Southern drawl: 'I got energy to burrrn'.

As with *La Ronde, Baby Doll* arrived in Britain with an already controversial status. In America, it was banned in many cities and publicly condemned by the Roman Catholic Censors, the National League of Decency. In Britain, its critical reception was somewhat divided. While certain critics were alarmed by a film they perceived to be 'revolting' and 'morally repellent', insisting that 'one "X" certificate hardly seems enough',[36] its reception was also favourable. The *Monthly Film Bulletin* considered *Baby Doll* to be 'an adult, perceptive and entirely individual film' while the *Daily Sketch* described it as 'a difficult and demanding film... one of the most serious made in America since the war'.[37] Further confirming its status as a 'difficult' and worthy film distinct from the usual 'commercial' fare, *Kine Weekly* later reported how *Baby Doll* 'pleased the critics, if not the masses'.[38] This begins to suggest why a film such as *Baby Doll* might be regarded as suitable for television coverage. While it may be perceived as 'revolting' and 'morally repellent', it could also be seen as 'difficult' and 'demanding'. Its director, Elia Kazan, had acquired a reputation as a realist director dealing with adult themes. When *Picture Parade* previewed his *A Face in the Crowd* (1957) (interestingly a damning satire on television), the Bond introduced the film by explaining: 'when Elia Kazan tackles any subject you may be sure that it will be treated realistically and dramatically - and that the subject itself will be provoking'.[39] After the preview of *Baby Doll* there followed a debate on the film chaired by filmmaker and critic Lindsay Anderson. Although few details are given in the script, it clearly explored the film's controversial status. As with the example of *Rashomon,* this suggests an attempt to deal with the film in a responsible and 'serious' manner. Here, however, its controversial nature is made explicit, and indeed partly constructed and circulated by television.

To take another example, *Picture Parade* also previewed *I'll Cry Tomorrow* which told the true story of Broadway star Lillian Roth's battle with alcoholism in the 1930s. Adapted from Roth's best-selling autobiography, the lead role was played by Susan Hayward who gave what many believe to be the best performance of her career. Advertised as a 'shocking and inspiring' story, Hayward offers a convincing performance as the alcoholic star, enabling the film to explore its subject in a frank, explicit and realistic manner. The film depicts the devastating impact of alcoholism on Roth's career, and ultimately ends with the singer facing her problem and accepting the help she needs. What might be perceived as a sensitive and explicit depiction of a serious social problem perhaps won favour with the BBC. However, although doing good business at the box office, the film was not received well by the British press. *I'll Cry Tomorrow* was variously described as 'a lurid journalistic account rather than the advertised "true story"' and a 'desperately embarrassing and tasteless film'.[40] Yet along with *Baby Doll*, it was clearly perceived as representing a suitable 'risk' for the BBC where pushing the boundaries of film content was concerned. The social significance of the genre, particularly films like *I'll Cry Tomorrow*, also reflected the Corporation's bid to educate and inform, as well as entertain. In sum, they could be accommodated within the broad concept of public service, particularly when the cinema programme was held in low critical regard.

In accordance with trends in film production and changing boundaries of film censorship, the BBC seem to have shifted slightly to a social, as well as artistic rationale for film regulation. It was now pictures such as *Baby Doll* and *I'll Cry Tomorrow* that were seen to represent 'exceptional merit'. There were also, however, similarities in the regulatory criteria of the BBC and the BBFC. If the British film industry felt that the censor showed a greater leniency toward the Continental film, a similar approach could be associated with the BBC. While genres such as horror were apparently regarded as outside of the BBC's selective rationale (for example, although promoting the film through other forms of coverage, the BBC decided against showing an excerpt from Hammer's X-rated sci-fi horror *X- the Unknown* [Leslie Norman,1956]), Continental X films seemed to pose no such problem. In fact, evidence suggests that television played a role in facilitating the growing acceptance of the Continental film into more mainstream British film culture at this time. Given that the Continental film is seen to epitomise all that was daring and different about the cinema by the late 1950s, this historical convergence between film and television is of considerable significance. It also contributes to an understanding of how foreign-language films (largely French and Italian) enjoyed an expanding circulation in British at this time. *Picture Parade*'s role in promoting these films needs to be understood within this wider context of their shifting exhibition and cultural consumption.

'Oh la la': The BBC and the 'Continentals'

In the 1950s British exhibitors turned to foreign-language films (and reissues) to fill their programmes in response to a product shortage from Hollywood. Described as the 'Continental invasion', this rise was reported in *Picturegoer* from the early part of the decade, and the magazine began its 'Continental Parade' in 1953. French films represented the largest part of this influx and in 1952 *Picture Post* claimed that in Britain as a whole, 491 cinemas now exhibited French films, four times as many as in 1950.[41] The trade press discussed and debated this shift and as *Kine Weekly* noted in 1957, for the 'average picturegoer... Italian and French offerings are no longer considered freaks. Now accepted at their true worth, Continental films have brought real money to the box office'.[42] Reporting on a

French film festival in London the *Daily Film Renter* discussed how: 'Time was when French films made their appeal to a restricted coterie of discriminating patrons, who saw in them an artistic approach to entertainment in which the mark of the director was of far greater importance than that of the stars',[43] and contrasted this situation with their now broadening appeal.

With the increasing popularisation of the Continental film, it was discourses on their 'risqué' nature which came to the fore and this can be seen in the way that the Continental film became synonymous with the X certificate. George Perry explains how, by the late 1950s, the X certificate was a 'badge of guaranteed salacity' (1977, p.189) and Anthony Aldgate describes its increasing association with the 'seamier realms of Continental production or the horror genre' (1995, p.14). There were, of course, different types of Continental film available to the British audience, as discussed in *Kine Weekly* by the Managing Director of a major distributor of Continental films in Britain. Entitled 'Can Continentals fill the gap?', the article explored whether the Continental film could ultimately appease the product shortage, and really make for British box office success. In discussing the range of Continental films it distinguished between the broad categories of the 'strong sex dramas' which often carry an X certificate and place the emphasis on dialogue rather than narrative, the visual comedy with an emphasis on farce and slapstick (such as the early films of Brigitte Bardot), and films with 'an emphasis on action rather than dialogue, a simple story, and better acting and higher production values to overcome the handi-cap [sic] of being foreign'.[44] The latter were seen to be epitomised by titles such as *Wages of Fear* (*La Salaire de la Peur*) (Henri-Georges Clouzot, 1953), *The Fiends* (*Les Diaboliques*) (Henri-Georges Clouzot, 1954) and *Rififi* (*Du Rififi Chez Les Hommes*) (Jules Dassin, 1955), and these films were repeatedly singled out as a 'turning point' in the British reception of the Continental film.

Wages of Fear, *Rififi* and *The Fiends* all received major circuit bookings, often sharing a bill with a British or American film. This was in contrast to the majority of the Continental films which were not generally shown on the main cinema chains. Particularly at a time of declining attendance, the screening of Continental films enabled smaller, independent cinemas to compete by offering something different. By the late 1950s many towns had a designated 'continental cinema'. This attests to the widening appeal of the Continental film, yet exhibition patterns (the reluctance to give them a circuit booking) equally suggest reservations on the part of the trade. This was perhaps due to the larger circuit's concerns over the commercial viability of X-rated films, but it also reflected deeper reservations about the popularity of the Continental film. In the trade press reviews it was only a minority of Continental films that were recommended for 'general bookings', as most were labelled for 'specialised' cinemas only. It is not entirely clear what is meant by this term - whether it simply means a 'Continental' cinema (some of which became colloquially known as 'the X-and-foreign-sex cinemas') - but the implication is that many were not perceived to be suitable for the general audience and that booking them was a commercial risk. *Kine Weekly* complained that the success of the Continental film was being hampered by over-cautious exhibitors,[45] but the reviews in the trade press can hardly have appeased their concerns.

Picture Parade expressed no such reservations, and it seems a crucial point that these films were previewed on television (now reaching an increasingly wide audience) at this time of their expanding circulation. It is difficult to reconcile some of the conservative reviews in the trade press with the coverage of Continental films on *Picture Parade*, as well as in *Picturegoer*'s 'Continental Parade'. Still Britain's most popular fan magazine, this was hardly addressed to a 'specialised audience', and the coverage of continental films was simply incorporated as part of the weekly release. The same was true of *Picture Parade*. The presenters Haigh and Bond would repeatedly stress how 'during the past few years Continental films have come into their own', or how 'Continental films - especially French - have been getting more and more popular lately...' This not only primed the viewer for the forthcoming preview but also attempted to disperse any lingering resistance to the Continental film, assuring the audience of their general appeal. That said, *Picture Parade* largely reviewed what the trade press described as the 'mainstream' or 'popular' Continentals. For example, the BBC seemed to be particularly fond of Brigitte Bardot's early comedies. In the introduction to *The Bride is too Beautiful* (*Bride La Mariee est trop belle*) (Pierre Gaspard-Huit, 1957), the presenter boasted that Bardot's 'name appeared more than any other' in their Continental spot, and along with Gina Lollobridgida, Sophia Loren and Martine Carol, she was one of the Continental stars which the trade perceived to have 'established star value'. While these stars certainly received extensive coverage in Britain, *Picturegoer* and *Picture Parade* featured a wider variety of Continental stars on a regular basis ranging from established names (such as Yves Montand and Simone Signoret) to lesser-known actors and actresses.

In *Picturegoer*, the female Continental stars were largely constructed through their sex appeal, and even *Picture Parade* indulged in some playful banter about 'sex-kitten' and 'gorgeous Pekinese', Brigitte Bardot. Yet like the majority of the 'mainstream' Continentals, many of her early comedies carried only an A or U certificate. While the 1950s Continental film may conjure up images of Bardot in *And God Created Woman* (*Et Dieu... crea la femme*) (Roger Vadim, 1956); her *Mam'selle Pigalle* (*Cette Sacree Gamine*) (Roger Vadim, 1956) (a farce in which Bardot plays a madcap schoolgirl whose father is being chased by the police) carried a U certificate, and was screened in a double bill with Disney's *The Great Locomotive Chase* (Francis D. Lyon, 1956). Her previous film *Mam'selle Striptease* (*En Effeuillant la Marguerite*) (Marc Allegret, 1955) was also regarded as somewhat tame and the *Daily Mirror* reported that 'it was not so saucy as it sounds', although *Films and Filming* suspected that it may have been 'toned down' for its British release.[46]

Perhaps, therefore, there was a conscious attempt by the BBC to preview not only the 'mainstream' Continentals, but also the more 'sanitised' examples. From the trade's point of view, these factors worked in dialogue given that 'strong sex dramas' were not considered suitable 'for general British booking'. However, *Picture Parade* also previewed some of the films with more 'risqué' connotations which indeed carried an X certificate. It is difficult to imagine the very middle-class tones of Peter Haigh introducing the saucy *Oh La La Cherie* (Pierre Gaspard-Huit, 1956) while explaining how he was 'always attracted to what the Continental cinema [has] to offer',[47] and the presenters often adopted a somewhat 'tactful' approach when describing the plots.

Certainly, despite the exploitation of the sexual explicitness of the Continental film, the BBC perhaps also valued them for their artistic associations. The trade press is evidence of this in stressing how they appealed to different audiences in different ways. For example, in reviewing the Italian *Woman of Rome* (*La Romana*) (Luigi Sampa, 1954) (starring Gina Lollobridgida), the *Daily Film Renter* describes it as a 'meaty drama for all types of specialist houses and the "X"-eries',[48] while with respect to *Card of Fate* (*Le Grand Jeu*) (Robert Siodmak, 1956), *Kine Weekly* complained that 'specialised audiences and sex seekers will be equally bored'.[49] The 'specialised' audience - the 'serious' Continental viewer - is here differentiated from the 'sex-seeker' or the 'X-erie', while simultaneously co-existing in the same space. This was perhaps one of the reasons why the BBC were more lenient with such products and it is notable that phrases such as 'delightful' or 'beautifully acted and directed' were often attached to the Continental films. Alternatively, the BBC were perhaps sometimes *unaware* of their certificates. If so, this suggests that while certain polices for film regulation had been established by this time, there was still a certain instability where their effective operation was concerned.

While the BBC had their own criteria where the regulation of film content was concerned, *Picture Parade*'s relations with 'adult' material shifted in accordance with changing production, distribution and exhibition patterns. In *Current Release* when the X certificate first appears we see only sporadic (and extreme) examples such as *Rashomon* and *La Ronde*. By 1956-58, however, we see an increasing number of social problem and Continental films, reflecting shifts in production and exhibition patterns. This reaches it's peak at the end of the decade when the X certificate moves into the mainstream of British film production.

'Are these the scenes we should "Parade"?': The British New Wave
The British film industry's hostility toward the BBFC reached its climax in 1958

when (after the BBFC made over fifty objections to the script of *No Trees in the Street* [1959]), the director, Jack Lee Thompson, launched an attack on the censor. Published in several journals and broadcast on BBC radio, this was entitled 'The Censor Needs a Change' and it claimed that 'British censorship [was] working against the filmmaker'.[50] Thompson not only targeted an inconsistent approach and the favouritism shown to foreign-language (and American) films. He also complained that in contrast to film, 'television can do anything it likes'. (He was referring to general television material here, such as plays). John Trevelyan, now secretary of the BBFC, responded to Thompson's complaints in *Films and Filming* and his reply was also broadcast on BBC radio. His response regarding television is particularly interesting as in order to support his case, he specifically drew on the context of its reception. He admitted that television had screened material that would not have been passed for cinema exhibition, but attempted to justify this with the claim that 'things have a lesser impact on the small television screen in the home than they would ... on the large screens of cinemas with a crowd of people producing a mass response' (quoted in Aldgate, 1995, pp.38-39). He continued that it was easier to 'switch off a television set' than to walk out of a crowded cinema, and this contrasts with contemporary standards (as well as the earlier complaints about *Rashomon*) in which television's accessible nature is deemed to require *stricter* regulation where films are concerned. Trevelyan emphasised that there had been 'a good deal of criticism' of some of the material presented on television and claimed that the BBFC were 'watching the position' and 'adjusting [their] policy accordingly'. This implies that, from the BBFC's point of view, what was permissible on the cinema screen was working less 'against' or in competition with television, than in *dialogue* with the new medium. Given that the BBFC were soon to demonstrate a more liberal approach, it can alternatively be seen that 'adult' films *were* constructed to compete with television, but not simply because they offered images and subjects unlikely to be seen in the home. According to Thompson it was because 'television can do anything its likes' and while this is overstating the case, this illustrates the fact that in many ways, films faced more stringent and established forms of regulation than their small screen rival.

Trevelyan was soon to be confronted with *Room at the Top* (Jack Clayton, 1959) which heralded the arrival of a new and exciting trend in filmmaking: the British 'New wave'. The history, aesthetics and implications of the British 'kitchen sink' films, so-called because of their depiction of the working class milieu, have been covered elsewhere and will not be reiterated here (see Hill, 1986, Murphy, 1992). These films dealt with sexual and social issues in a strikingly explicit manner, and my interest is in their impact on film censorship. John Trevelyan described *Room at the Top* as a 'milestone' in the history of the BBFC and although awarded an X certificate (for what Trevelyan described as 'rather more frankness about sexual relations in the dialogue than people had been used to') (Aldgate, 1995, p.33), the BBFC were congratulated for having the courage to pass the film at all. Due to its increasing association with the seamier realms of Continental cinema, the X certificate had fallen into disrepute. One critic commented that: 'at long last a British film which is truly adult. *Room at the Top* has an X certificate and deserves it - not for any cheap sensationalism but because it is an unblushingly frank portrayal of intimate human relationships' (Murphy, 1992, p.15). *Room at the Top* was seen as an opportunity for the rehabilitation of the X-rated film, the promise of which had fallen short of critical expectations. By the early 1960s (and the success of other New Wave films), the X certificate had proved its box office worth and it 'moved into the mainstream of British cinema', and as Dewe Mathews notes, 'now that the territory was commercially approved' even the Rank Organisation followed the trend with the X-rated *The Wild and the Willing* (Ralph Thomas, 1962) (1994, p.145).

This can thus be seen as a period when, partly due to the emergence of the 'kitchen sink' films, the X certificate became institutionalised. As a result, *Picture Parade* began to review X-rated films more frequently. This shift in film censorship was foregrounded in the first edition of the 'new' *Picture Parade* in 1959 (discussed in the next chapter), when John Trevelyan discussed the current situation and the difficulties involved in censoring different types of films. In introducing the feature the presenter, Robert Robinson noted that:

> It seems to me that the British film censor is getting more and more broadminded. Once upon a time he blushed very easily and assumed we did the same. But over the last year or two, he has become bolder and bolder. I don't think you could have seen this excerpt from *Look Back in Anger* [Tony Richardson, 1959] five years ago...

Picture Parade previewed many of the British New Wave films such as *The Entertainer* (Tony Richardson, 1960), *Saturday Night and Sunday Morning* (Karel Reisz, 1960) and *A Kind of Loving* (John Schlesinger, 1962) which, like *Look Back in Anger*, carried an X certificate. On one occasion the programme chaired a debate about film censorship when Robinson explained that *Saturday Night and Sunday Morning* had been 'banned unseen' by the Watch Committee of Hyde and Cheshire, and the interested parties were invited onto the programme to discuss the matter. This again indicates television playing a role in discussing shifts in film subject matter and censorship, while simultaneously providing an outlet for a consideration of its cultural implications.

These films were discussed by critics at the time as a 'renaissance' in British film production after the less acclaimed efforts of the previous decade. They were evidently seen as particularly worthy by the BBC which had always been keen to promote the achievements of British cinema. The 'kitchen sink' trend also permeated drama in the early 1960s, including television plays. When investigated by the Pilkington Committee the BBC defended their decision to screen a series of Sunday plays (which by their own admission 'contained an undue emphasis on the sordid and the sleazy') on the grounds that it was their 'duty' to reflect a 'significant element in current dramatic writing' (Shaw, 1999, p.80). Perhaps a similar perspective was taken on the films. Yet by the early 1960s it is harder to discern the BBC's criteria for regulating film content. The British New Wave may have been deemed a 'renaissance' in British film, but *Picture Parade*'s acerbic and blunt Robert Robinson seemed to have little interest in their aesthetic prestige. In one edition an excerpt from *Saturday Night and Sunday Morning* was followed by a clip from *Butterfield 8* (Daniel Mann, 1960), a film in which Elizabeth Taylor plays a prostitute. The film was regarded as very daring at the time and when introducing the excerpt Robinson claimed that:

> Sex seems to be pursuing me relentlessly in every film in this programme. It's hard to think of a more interesting subject, but you must blame the film industry if your diet [tonight] has been somewhat unvaried. *Butterfield 8* is the story of a nymphomaniac whose promiscuity brings her into contact with ... uninteresting men.[51]

Robinson censors the plot by describing Taylor as 'promiscuous' and unbalanced, rather than a prostitute, yet he nevertheless foregrounds the sexual connotations of these films. Perhaps in an attempt to deflect possible criticism, he also claims that the sexual emphasis is simply a reflection of current trends in film production, rather than a deliberate *selection* by the BBC. By the early 1960s members of the film industry argued that by showing such scenes, *Picture Parade* was a bad advertisement for the cinema. To take one example, an angry exhibitor claimed that it was 'calculated to drive people away from the cinema' and the article was entitled 'Are these the scenes we should 'Parade?'[52] He took particular offense to an edition including *Town Without Pity* (Gottried Reinhardt, 1961), *Homicidal* (William Castle, 1961) and an adaptation of Tennessee Williams' *Sweet Bird of Youth* (Richard Brooks, 1961) in which the viewer witnessed a 'bombardment' of images including a trial of rape, a decapitation, and 'offensive sexual morals'. This collection of scenes made the exhibitor feel that the 'future of the film industry was very dark indeed', and according to BBC Viewer Research reports, even members of the audience occasionally complained, criticising the amount of 'sex and violence' in the programme.[53] In a complete *reversal* of conventional arguments, far from an industry eager to exploit such representations in a bid to compete with the small screen, it is television which is associated with these controversial images, accused of giving the film industry a 'bad name'. This appears to be based on the ideal that the cinema is 'family entertainment', or that it should be, an ideal which by 1960, was clearly no longer realistic.

In the early 1960s under the new liberal direction of Hugh Greene, BBC programme makers were encouraged to push the boundaries of programme content, but many were critical of this new regime. It was not only contemporary trends in drama which attracted concern, as evidence of liberalism could be found more widely, ranging from the erotic dancing of 'Pan's People' (*Top of the Pops* [BBC, 1964-]) to the so-called 'smutty' jokes of the satirical *That Was The Week That Was* (BBC, 1962-63). By 1964 pressure groups mounted the 'Clean up TV' campaign while the press demanded to know 'What's Gone Wrong with the BBC?'[54] Looking back at the 1960s, Mary Whitehouse claims that the BBC was one of the most important forces in 'the creation of the permissive society... the voice of a new morality', although this is rather more sweeping than the BBC's own conception of its role as a 'mirror or transmitter' of the prevailing cultural climate (Shaw, 1999, p.82). *Picture Parade*, then, was part of this changing context, and its critics reflective of a wider chorus of disapproval. It is clear that shifts in film production and film censorship worked in dialogue with shifts in television programming, and a changing social climate, to produce the striking example of *Picture Parade* in the early 1960s. *Picture Parade* was apparently a programme where 'suggestive bedroom scenes' were not only likely but *expected,* and this returns us to where we began, with the *Sunday Independent*'s disgust at the series and its reflection of a general decline in the BBC's moral standards. This synergy is shaped by the fact that, in the 1960s, both film and television pushed against restriction, although this interaction clearly also had a history in the previous decade. Given its constant involvement with contemporary film culture, it is largely the cinema programme which insists that we modify our understanding of these relations. To put it simply, it seems unlikely that viewers saw 'adult' and controversial films as purely an attempt to compete with television when introduced to *Baby Doll, Oh La La Cherie, Saturday Night and Sunday Morning* (and even *Psycho* [Alfred Hitchcock, 1960]) by the BBC.

With regard to the cinema programme, the history of these previews is also the history of how the regulation of film developed on British television, and the aesthetic, institutional and cultural factors which surrounded it. While television was equally shaped by the more liberal climate of the 1960s, the attention paid to the regulation of television content only increased after this period, and in later years, this has been particularly so when it comes to the screening of feature films. This further emphasises this period as one of transition, and the context for surprising and challenging historical evidence. The image of the BBC boldly introducing *Rashomon* to a family audience acts as a tantalising indicator that the relations between film and television were far more complex (and critically interesting) than we have been led to believe.

It is certainly true that the cinema developed these strategies in part in a bid to compete with television, but this was also the time when television became a routine element of film promotion. The discussion in this chapter on both widescreen processes and X certificate films can be seen to support Christopher Anderson's argument that the 'boundaries that separate the media in our culture are the products of discourse, ... discourse generated by the media industries and ... by scholars and critics' (1994, p.14). The cinema programme indicates more widely a context in which the identities of film and television are not only historical, but are also fluid and permeable - operating in a shifting dialogue. If the cultural identities of cinema and television are historical and subject to constant change, it is perhaps no surprise that the 'world' of cinema and television explored in this book was subject to redevelopment as the 1950s came to a close.

Notes

1. *Kine Weekly*, 4 January, 1952, p. 2.

2. *Kine Weekly*, 17 December, 1953, p. 6

3. 13 May 1952. Notes on meeting between BBC and film industry to discuss *Current Release* programmes. T6/104/2.

4. Kine Weekly, 5 July. 1956, p. 28.

5. *Picture Parade* script no. 32.

6. *Picture Parade* script no.52

7. 8 May, 1957. Keith S. Allen to Alan Sleath. T6/405/1.

8. *Film Fanfare*, no.12.

9. *Today's Cinema*, 31 January, 1956: 3.

10. 29 November, 1957. Cecil Madden to British Lion. T6/360.

11. *Kine Weekly*, 13 December, 1956: 8.

12. *Picturegoer*, 7 March, 1953, p.7.

13. *The Advertiser*, 9 March, 1957, p. 12.

14. 22 April, 1962. BBC press cuttings, box PP62.

15. 'The Censor Talks About his X' in *Picturegoer*, 8 March, 1952, p.8.

16. *People*, 6 April, 1952.

17. 1 February, 1952. P.H Dorte to W. Farquarson-Small. T6/104/2 .

18. *Daily Film Renter*, 23 January, 1952, p.3.

19. *Kine Weekly*, 7 February, 1952, p.8.

20. BFI microfiche on *Rashomon*.

21. *Current Release* script, no.6

22. Undated letter from Gwen Collins, Hampstead, NW3, to BBFC. T6104/2 .

23. 1st April, 1952. Arthur Watkins to BBC. T6/104/2 .

24. 9 April, 1952. S.J de Lotbiniere to Arthur Watkins. T6/104/2 .

25. 4 April, 1952. Draft letter by Cecil Mc Givern. T6/104/2 .

26. Undated memo, 'BBC Policy on Films shown in Current Release'. T6/104/2.

27. *Picturegoer*, 8 March, 1952, p.8.

28. 29 August, 1952. BBC Press Cuttings, 1951-2.

29. 9 April, 1952. S.J de Lotbiniere to Mc Givern. T6/104/2 .

30. *Picturegoer*, 26 July, 1952, p.26.

31. 4 April, 1952. 'BBC Policy on films shown in Current Release'. T6/104/2 .

32. *Today's Cinema*, 19 December, 1957, p.10.

33. 7 August, 1956., '"A" certificate films', G.del Strother to Daphne Turrell. T16/543.

34. *Today's Cinema*, 8 August, 1956, p.12.

35. *Today's Cinema*, 7 January, 1958, p.3.

36. The *Times*, 24 December, 1956 and *Sunday Dispatch* 30 December, 1956. BFI microfiche on *Baby Doll*.

37. *Monthly Film Bulletin*, 20 December, 1956, p.14, and the *Daily Sketch* 20 November, 1956.

38. *Kine Weekly*, 12 December, 1957, p.7.

39. *Picture Parade* script no.52.

41. 'The French for Talent', *Picture Post*, 14 February, 1953, p.24.

42. *Kine Weekly*, 16 September, 1957: 3

43. *Daily Film Renter*, 26 March, 1957:6

44. 'Can Continentals Fill the Gap?', *Kine Weekly*, 5 July, 1956, p.25.

45. *Kine Weekly*, 5 July 1956, p.25.

46. *Daily Mirror*, 12 November, 1956, *Films and Filming* December, 1956, p.25.

47. *Picture Parade* script no.44.

48. *Daily Film Renter*, 22 March, 1957, p.3.

49. *Daily Film Renter*, 22 March, 1957, p.3, *Kine Weekly* 31 May, 1956, p.32.

50. *Films and Filming*, July, 1958, p.8.

51. *Picture Parade* script no.4.

52. *Today's Cinema*, 24 November, 1961. BBC press cuttings box PP62 .

53. 15 December, 1959. BBC Viewer Research Reports .

54. *Auntie: the Inside Story of the BBC: Making Waves 1960-70*, BBC1, 11 November, 1997.

Chapter Eight

'Picture Parade in Long Trousers':
Maturity and Change in the Cinema Programme

Current Release, Picture Parade and Film Fanfare emerge from a specific and unique period in the cultural co-existence of cinema and television, a context which provides much of their historical significance. The Introduction to the book quoted Picturegoer's perception in 1952 that the BBC's first cinema programme may represent the 'entertainment of the future'.[1] In many ways it did (the genre remains a constant presence in the schedules today), but what the magazine could not foresee was that the impending shifts in cinemagoing would render the scope of this 'synergy' less central than it seemed to believe. The genre was arguably at its height in the 1950s, endowed with a popularity and cultural significance it has not enjoyed since. This was shaped by the early competition between the channels for entertainment programming, yet the very reason that cinema featured so prominently here was because of its status as a popular mass interest. By the end of the decade these contexts were beginning to shift as both cinema and television looked toward changing cultural roles.

The late 1950s saw an institutionalisation of the relations between cinema and television where the cinema programme was concerned. Film review series continued to be produced after this time, existing within the institutional foundations constructed by their predecessors. Yet as an area of debate and dialogue between the two media, the importance of the cinema programme seemed to decline. This was in part due to the waning of its novelty factor - the genre had become conventionalised - but it seems to have been more than this. By late 1957 and early 1958 there were already signs of change, and the film industry's approach toward the programmes became particularly contradictory, impatient and critical. Still plagued by fears that the public would 'stay at home' to watch the programmes rather than attending the cinema, the film industry demanded that Picture Parade and Film Fanfare be reduced in length. They then complained that Film Fanfare was too brief and although the BBC initially refused to cut Picture Parade, when they did reduce its running time (possibly due more to programme planning and studio space than the film industry's requests), the KRS implored them to increase it once more.[2] In October 1957 Rank, Fox and Warner Bros (the three companies which, aside from MGM, contributed most to the cinema programmes) announced that they would no longer provide clips for Picture Parade. Picture Parade began with titles which claimed that it appeared with the 'full co-operation with the film industry', but the BBC lamented that the changes 'cannot help weakening the efficiency of Picture Parade which can no longer be called a magazine with "film industry co-operation" since 3 major companies are out'.[3] This increasing hostility was perhaps shaped by the number of feature films screened on television by this stage. As detailed in Chapter One, in 1956-57 a deal between the BBC and the British film industry failed and both channels eventually obtained films from elsewhere. (ABC acquired a package of Korda films, and the BBC announced its purchase of 100 films from RKO.) Scheduled at peak times on a Saturday night, the films were heavily promoted in the *TV* and *Radio Times*, and *Picturegoer* began its 'Home Screen Film Parade' in 1958, reviewing the films screened on television each week. The age of films on television had begun, and FIDO's defensive strategies had only prompted television to obtain films from elsewhere. Critics of the programmes might draw certain conclusions from this. Although they were not purchased from the British film industry, television had finally obtained attractive feature films. If the cinema programme was conceived as a 'bargaining tool', the reason for its existence had ceased.

While this is a selective and problematic conception of a genre which had its popular roots in broadcasting from the late 1920s, there was a sense in which the 'cinematic' sets, the abundance of stars, and the utter enthusiasm for the medium, represented the climax of the popular cinema programme on British television. *Film Fanfare* ceased in 1957 and *Picture Parade* in mid-1958. *Film Fanfare* was succeeded by ABC's *Box Office* (1957-8) which, not unlike its predecessors, brought news of current films and coverage of the stars. It also adopted the BBC's idea of constructing a 'realistic' interior, complete with cinema box office and cashier. However, it seemed to be a less ambitious programme than its predecessors and after encountering industry hostility, it ended in early 1958. *Close-up*, AR's star profile feature, continued, but no other cinema programmes appeared regularly on (London) ITV. The BBC also reduced the time devoted to film programming and the features that remained adopted a visibly different approach toward the cinema. At the end of *Picture Parade* in 1958 Haigh and Bond thanked the audience for 'having us in your homes',[4] and a key argument of this book is that this really marked the end of an era in which British film culture permeated the domestic sphere in quite the same way. This shift is best exemplified by the development of *Picture Parade*, which returned in 1959 and continued until 1962. When the programme re-emerged, the film industry and the viewers found it much changed. *Picture Parade* is particularly illustrative here as it provides a snapshot of the development of the cinema programme from the mid-1950s into the following decade, and enables a consideration of the various factors which shaped its changing form.

'Goodbye to the "fan-club" stuff': The 'New' *Picture Parade*

Upon its return, the BBC decided that *Picture Parade* required a 'fresh producer and a new personality'.[5] Former producer Alan Sleath was replaced by Richard Evans who worked closely with the Editor, Christopher Doll. This was not necessarily a drastic change as both Evans and Doll had previously directed the programme, and Doll acted as producer for a time when Sleath was absent. Sleath, however, had had a particular penchant for premieres and 'showbiz', and covered film events for the BBC outside of *Picture Parade*. In terms of presentation, it is clear that Haigh and Bond were no longer regarded as the most suitable choice. It has been suggested that Peter Haigh 'fitted perfectly into the television era of dinner jackets and clear enunciation', but that his career declined 'when the look of television began to change'[6]. The extremely polite and genteel address of television in this period is epitomised by the original *Picture Parade*, and the above comment ('when the look of television began to change') suggests that this was also associated with particular visual codes or signifiers. While it was *Film Fanfare*'s presenters which had worn dinner jackets and bow ties, this attire encapsulates key aspects of the visual and verbal address of 1950s television. When the first edition of the new *Picture Parade* aired in 1959,[7] viewers were not greeted by the smiling, mannered and amiable duo of Haigh and Bond, but by the acerbic, sarcastic and blunt Robert Robinson. Robinson had a wide journalistic background by this time. Among other positions, he had been TV columnist for the *Sunday Chronicle* (1952), film and theatre columnist for the *Sunday Graphic* (1956) and radio critic for the *Sunday Times* (1956). He later wrote a weekly column, 'Private View' in the *Sunday Times* (1962) and was film critic for the *Sunday Telegraph* (1965). He had already made his debut as a TV presenter prior to *Picture Parade* on the arts programme *Monitor* (BBC), and in the 1960s he presented programmes such as *Points of View*, *The Look of the Week*, and *Call My Bluff*. (Although appearing less frequently, he remained active in television throughout the 1970s and 1980s) [8]

Aside from his more 'professional' journalistic status, Robinson was also situated in a different studio context. Not unlike many real venues, when *Picture Parade* ended in 1958 it closed the doors of its 'super-cinema' for good. Upon its return, it had little interest in such an exuberant aesthetic. Rather than a palatial foyer, Hollywood desk and cinematic viewing theatre, the new series favoured a more functional approach and simply used a couch, armchairs and a coffee table. The programme continued to interview stars and directors and, of course, review popular films, but the stars ceased to dominate the programme in the same way. The BBC's Roving Eye camera was no longer peeping in on premieres or, in the style of *Film Fanfare*, searching out a visiting star as they arrived at the airport. Much of the star coverage was by way of straightforward interviews, and the construction of the stars is a crucial indicator in considering how *Picture Parade* had changed.

In contrast to Haigh and Bond's reverence and affection for the stars, Robinson was rather more sceptical of these fabulous ideals. One of his opening comments in the new series was about Elizabeth Taylor:

It's usually a safe bet that the real salary of any film star is the cube root of the figure you read in the gossip columns, but I have it from a man who looked me straight in the eye that Elizabeth Taylor is getting one million dollars for her performance as Cleopatra. If she lives simply, it should be sufficient.[9]

A similar scepticism and sarcasm pervaded his more general discussion of the stars, and after interviewing Sylvia Syms in connection with her role in *Express Bongo* (Val Guest, 1959), Robinson thanked the actress before addressing the viewer with the comment: 'It's always interesting to find out whether film stars are flesh and blood'.[10] Here Robinson implies that film stars may be phoney and artificial, a mere media construction, and this was in stark contrast to the sincerity of Haigh and Bond and their expression of a simultaneous deference and friendliness toward the star. Perhaps most interestingly, while (in drawing on traditional myths of fame), the earlier *Picture Parade* and *Film Fanfare* had talked excitedly about the 'magic' of 'star quality' and 'talent', Robinson even dared to suggest that film stars were not 'exceptional' at all:

I daresay any of us could name twelve girls capable of acting Brigitte Bardot into a cocked hat. I daresay any of us could name twelve others whose figures are just as good... What makes Bardot so different? Well perhaps she isn't different. Perhaps that's the secret....[11]

This was very different from *Picture Parade*'s previous 'Continental Cinema' spot in which Haigh and Bond were only too eager to welcome the star they called 'the sex kitten' or 'gorgeous pekinese' as she flirted her way through (what they termed) 'delightful' French comedies.

Although appearing much less frequently in the later series, the 'behind-the-scenes' coverage also came under Robinson's critical eye. In one edition he nonchalantly explained how:

a company making a large and expensive film often makes another film about the film. They shoot the cast and the technicians at work and everyone does an awful lot of smiling to make it perfectly clear what a cheerful bunch they are, and the result is used as publicity for the real film.[12]

Rather than offering the viewer a privileged glimpse of activity at the Warner Bros studios, or closer to home at Elstree or Pinewood, Robinson distances himself from the coverage by criticising its promotional rhetoric - and simultaneously foregrounds its construction of a 'phoney' perspective on filmmaking. Sarcasm and cynicism pervaded his commentary more generally and while this had particular implications for the coverage of film, it also reflected on the wider popularity of satire at this time. As Andrew Crisell explains in his discussion of the BBC's satirical and topical *That Was The Week That Was* (aka *TW3*) (1962-3), various factors contributed to this trend including increased media consumption, rebellious youth culture, and a mood of political scepticism (1991, p.146-7). Emerging at the end of the 1950s and early 1960s, theatre revues and magazines contributed to a satire boom and television's primary manifestation of this began in 1962 with the emergence of *TW3*. Crisell's description of the historical significance of *TW3* elucidates the shifts in the cinema programme at this time:

> [TW3] marked television's coming-of-age in the sense that many of its jokes and references depended for their effect on the audience's knowledge of a culture which had largely been created... by the medium itself. That such knowledge could be assumed is a measure of the extent to which television had embedded itself in the national consciousness

(1997, p.119).

Robinson's comments, jokes and references indeed depended for their effect on 'the audience's knowledge of a culture which had largely been created by the medium itself' or more specifically, the televised cinema programme. Traditional theories of generic evolution hold that a genre begins in a naive state, evolving toward a greater awareness of its own myths and conventions (Shattuc, 1997, p.138). When situated within the broader historical context of satire, Robinson's approach seems to bear this out. His comments about the stars or the 'behind-the-scenes' perspective implicitly traded upon a knowledge of the earlier series and their particular approach toward, and relations with film culture. Although more likely shaped by Robinson's own personality than a deliberate intention by the BBC, the later *Picture Parade* seemed, at times, to be almost making fun of the earlier programmes, having a laugh at their expense. The satire boom was largely the creation of Oxford and Cambridge graduates (Crisell, 1997, p.150) and it is perhaps significant that, following the first edition of *Picture Parade* in 1959, a warden at Oxford University claimed that he was 'going to recommend it to his students'.[13]

The film industry still resisted the adoption of an outright critical approach, and Robinson recalls how the BBC instructed him to provide 'a little perspective - not too much... because the show was to be informative not critical' (Robinson, 1997, p.136). Although his role was essentially to offer a plot synopsis of the films, Robinson explains how he soon discovered that there was more than one way of doing this (p.140). While not openly critical of the films, Robinson's cynicism foregrounded, for example, implausible plots, characters or situations. In fact, his approach now looks very much like a precursor to the rapid and knowing patter of the BBC's Jonathan Ross on their current *Film 2005*, and has more in common with this contemporary example than previous presenters of the genre. After the first edition of the new series the *Daily Mail* noted approvingly that 'The BBC rightly assumes that there is room for a reasonable critical programme about the film industry',[14] one of few positive appraisals of the cinema programme since its inception. The *News Chronicle* ran the headline '*Picture Parade* in long trousers' and claimed that it had done a remarkable bit of growing up. Gone is the "fan-club" stuff with television fawning humbly at the knees of mother cinema. In its place is a knowledgeable inquisitiveness and a touch of asperity'.[15]

The intention here is not to provide a neat homology between the changing role of the cinema and its 'reflection' on television. This relationship, we have seen, is one of active construction shaped by a myriad of different factors. Rather, this chapter considers various discourses which may have *contributed* to the shifting role, appearance and status of the cinema programme as a new decade dawned.

Arguably, one of the most important factors here was the decline of the cinema as a mass medium. In 1950 yearly admissions amounted to 1396 million, by 1960 this had fallen to 515 million (Spraos, 1962, p.14). Admission figures from the Board of Trade indicate that the steepest decline occurred between 1957 and 1959 (Spraos, p.24) - interestingly exactly the point when the 'new' *Picture Parade* began. When it returned in 1959 it was scheduled on a fortnightly, rather than weekly basis. It had been reduced to half an hour in length, and was screened at the later time of 10:30-11:00pm. It was not so much that a weekly review was no longer deemed necessary, but the programme was clearly given less centrality in the television schedule. It would certainly seem significant that by 1962, *Picture Parade* had become a *monthly* feature. Much to the annoyance of the trade press, in one edition Robinson inquired, 'Do *you* still go to the cinema?', and the fact that this question could even be asked at all is significant. It indicated a shared knowledge and understanding that the social role of the cinema was changing, had already changed, a process of which the cinema programme was now only too aware.

Addressing 'Intelligent Audiences': Into the 1960s

A clear mark of the cinema's mass appeal and central place in British leisure culture was the circulation of the fan magazine, and this is precisely the point when *Picturegoer* (closely followed by its rival *Pictureshow*) came to an end. In pursuit of the youth audience Britain's most popular film fan magazine expanded into other areas and, emphasising the increasing importance of pop music, in 1960 it merged with *Disc Parade*. Soon after this time it announced its transformation into *Date Girl*, 'the magazine for smart young women', which struggled to build a readership and ceased soon after (Baker, 1985, p.210). Christine Geraghty claims that *Picturegoer* was overtaken by 'new forms of popular culture in which the cinema began to look rather old-fashioned' (2000, p.11). It is doubtful that the cinema (with its continuing technological innovations and increasingly controversial content) was perceived to be simply 'old-fashioned' by teenagers, now the mainstay of its audience. Yet as film increasingly jostled for space with television and pop music, Geraghty is right in suggesting that within *Picturegoer* itself, the cinema seemed outdated as the magazine struggled to address its younger audience. The book has touched on the ways in which the magazine had flourished in a context in which British film culture pervaded everyday life. It traded on an implicitly shared knowledge of this context in which the stars were seen as 'friends', and the British film industry was affectionately referred to as 'British filmdom'. Once the family audience began to move away from filmgoing, and thus from *Picturegoer*, this dialogue with the magazine's readership began to wane (Holmes, 1998).

This shift can be discerned in other areas. In 1961 Roger Manvell claimed that 'the large unthinking audience has transferred the major part of its habitual attention to television' which meant that cinema 'could have some success with films that appeal to the intelligence of audiences' (cited in Geraghty, 2000, p.13). The end of the 1950s looked forward to the next decade when there was an increasing emergence of a more academic discourse on film. Surveying the 1960s, James Quinn's *The Film and Television as an Aspect of European Culture* (1968) asserted that 'During the last ten years there has been a marked increase in the number of people, including a high proportion of undergraduates and students of all kinds, taking a deeper interest in the cinema' (p.52). While still very much in its infancy, the academic study of film began in British universities in 1960. Equally, the developing role of the British Film Institute is important here in formalising notions of 'film education' (see McArthur, 2001), and could usefully be considered much more closely in terms of its relations with television's coverage of film at this time. While it is again clear that these strategies were very much in their early days, schools and colleges initiated film appreciation, taking an increased interest in its educational value (Quinn, 1968, p.55). The year that *Picture Parade* re-emerged marked the advent of the BBC's school's programme *Looking at Films*. Produced in conjunction with the BFI, the series was intended to give 'young people a basis on which to build critical appreciation and to show some of the processes that go into making a film'.[16] However loosely defined, the bid to foster film education had permeated the BBC's programming from the earliest days of radio, and produced as part of the educational school programmes, *Looking at Films* represents a consolidation of this approach. While 1959-62 is still early to speak of the institutionalisation of an academic approach to film, it was at this time that journals such as *Cahiers du Cinéma* and *Movie* developed an 'auteurist' approach to film which, in claiming the director as the locus of critical value, raised the status of film as an art form. By the end of the 1950s journals such as *Films and Filming* were indeed pushing for film to be conceived as an art form for adults (Geraghty, 2000, p.10) and along with *Sight and Sound*, they enjoyed an expanding readership.

Just before *Picture Parade* reappeared in 1959, the BBC perhaps attempted to address the 'intelligent audiences' conceived by Manvell with their series *The Cinema Today*.

Launched at the beginning of the year and accompanied by an article in the *Radio Times* by Dilys Powell, *The Cinema Today* was devised by British director and writer Lindsay Anderson. Although *The Cinema Today* included clips from contemporary films, it was clearly not a review feature. The first few editions were organised around the theme of national cinemas and beginning with the example of Italy, it went on to explore Scandinavian, British, American and Russian cinema. The *Times* noted that *The Cinema Today* was aimed at a more minority audience than *Picture Parade* and described its appeal to 'the serious student of film'.[17] The *Daily Worker* was impressed by the programme's international flavour, but lamented that the majority of its films 'would never come any nearer to British film lovers' than London's NFT.[18] This link was explicitly part of the BBC's project as the NFT agreed to screen certain films to complement the programme. This is prefigured by the occasional example in the earlier *Picture Parade*, although with respect to *The Cinema Today*, the NFT were explicitly organising their programme around the television coverage.

Emerging in 1964, one of the most significant cinema programmes of the decade was undoubtedly Granada TV's *Cinema*. When the programme was first conceived Granada insisted that 'It should not be a movie magazine, nor simply a show that boosted new releases. It should instead embody a *serious popular approach* to films [my emphasis]'.[19] This makes clear that the discursive and textual boundaries of the cinema programme were being re-drawn: within the 'popular' (and mass) address of television, there was space for a 'serious' approach to the cinema conceived as worthy of critical attention. In the *ITV Handbook* (1965), *Cinema* did not appear under 'light entertainment' (as had been the case with other film features), but was listed under 'The Arts'. It is not, however, simply a case of the changing role of the cinema impacting upon television's coverage of film, and it is not my intention to offer a simplistic or reflectionist analysis of this shift. The greater complexity of the situation is clearly emphasised by Quinn's study, and his observations about television are worth quoting at length:

> There appears to be general acceptance of the fact that there has been some change in attitude during the pást 10 years. There also seems to be a large measure of agreement... as to the reasons for the change. The most significant of these is television. Both the BBC and Independent Television companies transmit programmes in which extracts from current and older films are discussed. At the popular level the Granada... programme *Cinema* commands a vast audience... Both BBC1 and [ITV] also televise programmes which include interviews with film directors and producers and BBC2 has screened a considerable number of foreign-language films accompanied by a critical assessment of their artistic significance
>
> *(1968, p.61).*

Quinn's argument, then, is that *television* played a not inconsiderable role in helping to reshape the changing cultural identity of cinema. BBC2 began in 1964 and is described by Quinn as the channel for 'the discriminating viewer' (Quinn, 1968, p.68), and it did much to stimulate a critical appreciation of film in the 1960s. The screening of feature films is also important. By 1960 the television schedules were becoming (what was then perceived as) saturated with films. Critics complained that 'Television has become a recording machine rather than a live medium'[20] and felt that it was developing into 'one massive ... movie museum'. These screenings also suggested, however, an audience becoming steeped in film culture in a readily accessible form. As the *Daily Telegraph* later suggested, 'television has filled so many empty hours with ancient films that it has created an audience well informed on the history of the cinema...'[21] Quinn's conception of the cinema programme as educational, however, is crucial. While the features of the 1950s may also have been 'dazzled' by glamour, *really* more interested in showcasing the latest 'charming' star or busily whipping through studio 'news', this process of education begins with them (or indeed with radio decades before). Quinn's argument challenges the conception that television's coverage of film culture - particularly the film review - has done little to stimulate a critical awareness of the medium. Quinn claimed that it was not television, but the film industry which had no 'genuine concern to promote a serious interest in the cinema' (p.59). What is perhaps important is that such a 'serious interest' could now be conceived at all.

That said, it is important not to over-emphasise this shift. Particularly for those who attended regularly, the cinema remained a popular medium, and the function of *Picture Parade* was to preview the latest releases, whether *Saturday Night and Sunday Morning* (1960) or *Carry on Nurse* (Gerald Thomas, 1959). Nor is it my intention to suggest that the new *Picture Parade* had nothing in common with its predecessors. However, the *Picture Parade* of 1959 did seem - however elusively - to be in some way shaped by the changing cultural role and conception of the cinema, and looked forward to some of the shifts outlined, even if it was not yet fully aware of them. While television was an important factor in prompting the decline of the cinema as a mass medium (thus contributing to its changing role), the cinema programme, and television's other coverage of film, perhaps played a key role in reshaping its shifting cultural identity.

The changing nature of the cinema programme was also influenced by the institutional structure of broadcasting and television's development into a mass medium. The programmes in 1956-58 were very much a product of the early competition between the channels, and what better way to compete for audiences than through coverage of the still mass interest of the cinema? By the end of *Picture Parade* in 1962, the channels were beginning to look ahead to more of a duopoly, and the earlier 'showbiz' aesthetic would itself have looked out of place in what was now a more mature medium. As indicated with respect to the discussion of Haigh and Robinson, this shift was also bound up not only with television's changing modes of address but also, perhaps, its attitude toward its subject. It is not simply a genteel politeness which characterises early programming but a certain innocence. This was related to its status as a new technology and shifted as the medium developed. In conjunction with the changing cultural climate of the following decade, Robinson addressed a much more 'knowing' television viewer, one at ease with the codes and conventions of television, and (given his approach), perhaps enjoyed seeing these codes subverted and undermined. Nevertheless, while after the first edition of the new *Picture Parade* in 1959 many viewers described it as 'adult' and 'informative', BBC Viewer Research also revealed that 'It was something of a disappointment to many'.[22] There were complaints about the programme's late scheduling and some felt that Robinson lacked 'personality' and was 'far too serious in his approach'. There was also the feeling that since the departure of Haigh and Bond, the programme had 'lost much of its vitality and was less enjoyable'. From a retrospective position, it is difficult not to agree with these claims. Television's rise and the cinema's decline overlapped just long enough to offer a startling, unique and infectious celebration of film culture which could no longer be sustained as the 1950s drew to a close.

The primary historical and critical value of the early cinema programme is undoubtedly its complication of the over-familiar and generalising picture of the period as one in which the cinema 'lost' its mass audience to television. Audiences *did* desert the cinema for television (and other leisure pursuits), but this shift was far from simple, uniform and straightforward. Within the current academic and cultural context and its questioning of grand, coherent narratives, it should come as no surprise that history refuses to conform to neat trends. The unsatisfying statistical sketch often used to characterise this shift is unable to address the changing experiences of the media, but also the ways in which they were interlinked. The reassessment enabled by the cinema programme (as well as my experience researching it) demonstrates the value, possibilities and pleasures involved in combining not only the history of film and television, but their respective disciplinary approaches. Yet I am not convinced that it offers any clearer answers to the questions of audience experience. The ways in which the reception, pleasures and uses of cinema and television shifted and interacted at this crucial point of change is fascinating. While audiences (particularly historical audiences) can remain typically elusive from a critical point of view (Ang, 1991, Jenkins, 2000), this elusiveness is also shaped by the cinema programme's status as a product of contradictory relations between film and television: it looked two ways at once. It is still possible to conceive in the 1950s cinema programme the sort of cinemagoing culture celebrated by *Picturegoer* - the 'knowingness' with which it constructed British film culture, the chance to win cinema tickets to the 'local' ABC, the attempt to satisfy the family audience, the tireless stream of current releases, stars and studio news. Viewing the programmes today, they still retain the vibrancy with which this approach was intended, although through a retrospective gaze, this is tinged with a certain parochial, although deeply appealing, insularity. Yet the cinema programme was also a site upon which British television planted more and more images of film and cinema in the domestic sphere. In bringing 'the world of the cinema to your home', it helped to shape the acceptance of, and familiarisation with, the domestic consumption of film.

If a sense of regret tinges my discussion of the transformation of the cinema programme at the end of the 1950s (and it does), this only serves to mark the extent to which the cinema files at the BBC's Written Archive Centre opened up a vast critical and historical space of film and television culture startling in its scope, implications and critical complexity. Television - like history - is culturally marked as ephemeral and transient, and this takes on a further significance in reconstructing the cinema programmme which epitomised a particular period of television, and revelled in a particular conception of the cinema, which were both disappearing as fast as the flickering images flashed across the screen.

Notes

1. *Picturegoer*, 9 February, 1952, p.8.

2. *Daily Film Renter*, 8 March, 1957, p.4.

3. 25 October, 1957 Madden to KRS T6/360.

4. *Picture Parade* script, 69b.

5. 30 September, 1958. Madden to ABPC T6/360.

6. Obituary for Peter Haigh, *Daily Telegraph*, 21 January, 2001.

7. The new *Picture Parade* began on 20 November 1959. Unlike the previous programme, this series did not run continuously, but was divided into three. I have used the letters 'a', 'b' and 'c' to indicate this when referencing the number of the script.

8. *Who's Who*, 1999, p.445.

9. *Picture Parade* script 1a.

10. *Picture Parade* script 2a.

11. *Picture Parade* script 7a.

12. *Picture Parade* script 4b.

13. 24 November, 1959. Kenneth Adam to Cecil Madden. T6/408/1.

14. *Daily Mail*, 24 November, 1959. BFI microfiche on *Picture Parade*.

15. *News Chronicle*, 8 December, 1959. BFI microfiche on *Picture Parade*.

16. 6 March, 1959. Gordon Smith to KRS. T6/360.

17. *The Times*, 7 February, 1962. BFI microfiche on *Picture Parade*.

18. *Daily Worker*, 13 March, 1959. BBC press cuttings box P669.

19. Quoted in *Year Ten: Granada TV* p.13

20. 4 January, 1962. *Bournemouth Evening Echo*. ITA press cuttings.

21. *The Daily Telegraph*, 3 September, 1966.

22. BBC Viewer Research report on *Picture Parade*, 15 December, 1959.

Bibliography of Sources

Primary Sources at the BBC Written Archive Centre

5 files of memos and correspondence on Film Talks on radio (1929-1953), R51/173/1-R51/173/5.

3 files of memos and correspondence on the radio programme *Picture Parade* (1945-49), R19/913/1- R19/915/1.

3 files of memos and correspondence on *Current Release* (1952-3), T6/104/1- T6/104/3.

15 files of memos and correspondence on *Picture Parade* (1956-62), T6/399/1-T6/414/1

Picture Parade scripts (radio) (1946-49) [on microfiche].

Movie-Go-Round scripts (radio) (1956) [on microfilm].

Current Release scripts (1952-53) [on microfilm].

Picture Parade scripts (1956-62) [on microfilm].

Magazines / Trade Press

Daily Film Renter (1952-1953, 1956-1958).

Picture Post (1952-1958) (selected issues).

Kine Weekly (1952-1953, 1956-1958) (selected issues).

Picturegoer fan magazine (1949-1959).

Radio Times (1952-1962) (selected issues).

Today's Cinema (1952-1953, 1956-1958) (selected issues).

TV Times (1955-1962) (selected issues).

Archival audio-visual material

12 editions of *Film Fanfare* (1956-7) held at the Associated British Pathé archives at Pinewood studios, Ivor Heath, Buckinghamshire.

1 edition (tx 7 August, 1956) of *Picture Parade* from the BBC's audio-visual archives in Middlesex.

Works Cited

Aldgate, Anthony (1995) *Censorship and the Permissive Society: British Cinema and Theatre: 1955-65* (Oxford, Clarendon)

Alexander, Karen (1991), 'Fatal Beauties: Black Women in Hollywood', in Christine Gledhill (ed.), *Stardom: Industry of Desire* (London, Routledge), pp.45-57.

Allan, Graham and Crow, Graham (1991), 'Privatization, Home-centredness and Leisure', *Leisure Studies* 10:1, pp.19-33.

Allen, Jeanne (1980), 'The film viewer as consumer', *Quarterly Review of Film Studies* 5, 4, pp.481-99.

Anderson, Christopher (1991), 'Hollywood in the Home: TV and the End of the Studio System', in Patrick Bratlinger and James Naremore (eds.), *Modernity and Mass Culture* (Indiana, Indiana University Press) pp.80-103.

Anderson, Christopher (1994), *Hollywood TV: The Studio System in the Fifties* (Austin, University of Texas Press).

Anderson, Janice (1991), *Marilyn Monroe* (London, Hamlyn).

Ang, Ien (1991), *Desperately Seeking the Audience* (London, Routledge).

Aumont, Jacques, Bergala A, Maries M, and Vernet M (1992), *Aesthetics of Film* (Austin, University of Texas Press).

Babington, Bruce (2002) (ed.) *British Stars and Stardom: From Alma Taylor to Sean Connery* (Manchester, MUP).

Baker, Bob (1985), 'Picturegoes', *Sight and Sound,* Summer, pp.206-9.

Balio, Tino (1985) *The American Film Industry* (Madison, University of Wisconsin Press)

Balio, Tino (1990) (ed.), *Hollywood in the Age of Television* (Boston, Unwin Hyman).

Barr, Charles (1986), 'Broadcasting and Cinema: 2: Screens within Screens', in Charles Barr (ed.), *All Our Yesterdays: 90 Years of British Cinema* (London, BFI) pp.206-225.

Barr, Charles (1996), 'They Think it's All Over: The Dramatic Legacy of Live Television', in John Hill and Martin Mc Loone (eds.), *Big Picture, Small Screen: The Relations Between Film and Television* (Luton, University of Luton Press) pp.47-76.

Beadle, Gerald (1963), *Television: A Critical Review* (London, Allen and Unwin).

Belton, John (1992), *Widescreen Cinema* (London, Harvard University Press).

Benjamin, Walter (1973) 'The Work of Art in the Age of Mechanical Reproduction', in Hannah Arendt (ed.), *Illuminations* (London, Fontana Press) pp.211-245.

Black, Peter (1977), *The Mirror in the Corner: People's Television* (London, Hutchinson).

Boddy, William (1993), *Fifties Television: The Industry and its Critics* (Urbana and Chicago, University of Illinois Press).

Briggs, Asa (1965), *The Golden Age of Wireless: The History of Broadcasting in the United Kingdom: Volume II* (Oxford, Oxford University Press).

Briggs, Asa (1979), *Sound and Vision: The History of Broadcasting in the United Kingdom: Volume IV* (Oxford, Oxford University Press).

Briggs, Asa (1995), *The History of Broadcasting in the UK: Volume V* (Oxford, Oxford University Press).

Brown, Geoff (1997), 'Paradise found and lost: the course of British realism', in Robert Murphy (ed.), *The British Cinema Book* (London, BFI).

Brown, Ivor (1952), 'Television in the Englishman's Castle', in *The BBC Yearbook of 1952,* pp.17-19.

Burton, Alan and Chibnall Steve (1999), 'Promotional activities and Showmanship in British film exhibition', *Journal of Popular British Cinema* (2), pp.83-99.

Buscombe, Ed (1971), *Screen Pamphlet 1: Films on TV* (London, BFI).

Buscombe, Ed (1991), 'All Bark and no Bite: The Film Industry's Response to Television', in John Corner (ed.), *Popular Television in Britain: Studies in Cultural History* (London, BFI), pp.197-209.

Bussell, Jan (1952), *The Art of Television* (London, Faber & Faber).

Butler, Jeremy G (1991) (ed.), *Star Texts: Image and Performance in Film and Television* (Detroit, Waynes State University Press).

Caughie, John (1991a), 'Before the Golden Age: Early Television Drama', in John Corner (ed.), *Popular Television in Britain: Studies in Cultural History* (London, BFI) pp.22-42.

Caughie, John (1991b). 'Adorno's reproach: repetition, difference and television genre', *Screen,* 32:2, Summer, pp.127-153.

Caughie, John (2000), *Television Drama: Realism, Modernism and British Culture* (Oxford, Oxford University Press).

Caughie, John and Rockett, Kevin (1996), *The Companion to British and Irish Cinema* (London, BFI Cassell).

Chapman, James (1998) 'Our finest hour revisited: The Second World War in British feature films since 1945', *Journal of Popular British Cinema* 1, pp.62-73.

Chapman, James and Christine Geraghty (2001) 'Editor's Introduction', *Journal of Popular British Cinema* (British Film Culture and Criticism), 4, pp.2-5.

Clarke, John and Critcher, Chas (1985), *The Devil Makes Work: Leisure in Capitalist Britain* (London, Macmillan).

Clarke, Michael (1952), 'Television Prospect', in Roger Manvell (ed.), *The Cinema 1952* (London, Penguin Books) pp.174-187.

Considine, Shaun (1989), *Bette and Joan: The Divine Feud* (London, Sphere Books).

Cook, Pam (1996), 'Introduction', in Pam Cook (ed.), *Gainsborough Pictures* (London, Cassell), pp.1-12.

Corner, John (1991), 'General Introduction: Television and British Society in the 1950s', in John Corner (ed.), *Popular Television in Britain: Studies in Cultural History* (London, BFI) pp.1-22.

Corner, John (1992) 'Presumption as Theory: "Realism" in Television Studies', *Screen* 33:1, Spring, pp.97-102.

Crisell, Andrew (1991), 'Filth, Sedition and Blasphemy: The Rise and Fall of Television Satire', in John Corner (ed.), *Popular Television in Britain: Studies in Cultural History* (London, BFI), pp.145-158.

Crisell, Andrew (1997), *An Introductory History of British Broadcasting* (London, Routledge).

Crow, Duncan (1957), 'From Screen to Screen: Cinema Films on Television', *Sight and Sound* Autumn, pp.61-64.

De Cordova, Richard (1990), *Picture Personalities: The Emergence of the Star System in America* (Urbana and Chicago, University of Illinois Press).

Demant, Canon (1954), 'The Unintentional Influences of Television', *BBC Quarterly* IX: 3 pp.136-142.

Desjardins, Mary (2002) 'Maureen O'Hara's "Confidential" Life: Recycling Stars through gossip and moral biography', in Janet Thumim (ed.), *Small Screen, Big Ideas: Television in the 1950s* (London, I.B Tauris), pp.118-130.

Dewe Mathews, Tom (1994) *Censored: The Story of Film Censorship in Britain* (London, Chatto and Windus).

Dimmock, Peter (1952), 'Television Out and About', in *The BBC Yearbook of 1952* pp.50-51.

Doane, Mary Ann (1989), 'The economy of Desire: the commodity form in/of the cinema', *Quarterly Review of Film and Video,* 11 pp.23-33.

Doherty, Thomas (1988) *Teenagers and Teenpics: The Juvenilization of American Movies in the 1950s* (Philadelphia, Temple UP)

Dovey, Jon (2000), *Freakshow: First Person Media and Factual Television* (London, Pluto).

Dyer, Richard (1987), *Heavenly Bodies: Film Stars and Society* (London, Macmillan).

Dyer, Richard (1998), *Stars* (New Edition) (London, BFI).

Eckert, Charles (1978) 'The Carole Lombard in Macy's Window', *Quarterly Review of Film Studies* 3 (1), Winter.

Epstein, Rebecca (2000) 'Sharon Stone in a Gap Turtleneck', in David Desser (ed.), *Hollywood Goes Shopping* (Minneapolis, University of Minneapolis Press), pp.44-63.

Ellis, John (1992), *Visible Fictions* (Revised Edition) (London, Routledge).

Ellis, John (1996), 'The Quality Film Adventure: British Critics and the Cinema, 1942-48', in Andrew Higson (ed.), *Dissolving Views: Key Writings on British Cinema* (London, Cassell), pp.66-93.

Ellis, John (2000), *Seeing Things: Television in the Age of Uncertainty* (London, I.B Tauris).

Evans, Jeff (1995), *The Guinness Television Encyclopedia* (Bath, Bath Press).

Eyles, Allen (1993), *ABC: The First Name in Entertainment* (London, BFI).

Eyles, Allen (1996), *British Gaumont Cinemas* (London, BFI).

Feuer, Jane (1992), 'Genre Studies and Television', in Robert C. Allen (ed.), *Channels of Discourse: Reassembled* (Carolina, University of North Carolina Press), pp.138-161.

Fiske, John (1987), *Television Culture* (Routledge, London).

Gaines, Jane (1989), 'The *Queen Christina* Tie-ups: convergence of show window and screen', *Quarterly Review of Film and Video* (11), pp.35-60.

Gaines Jane, and Herzog, Charlotte (1990) (eds), *Fabrications: costume and the female body* (London, Routledge).

Gamson, Joshua (1994) *Claims to Fame: Celebrity in Contemporary America* (Berkely, University of California Press).

Gamson, Joshua (2001) 'The Assembly Line of Greatness: Celebrity in Twentieth-Century America', in C. Lee Harrington and Denise D. Bielby (eds.), *Popular Culture: Production and Consumption* (Oxford, Blackwell), pp.259-282.

Geraghty, Christine (2000a), *British Cinema in the Fifties: Gender, Genre and the 'New Look'* (London, Routledge).

Geraghty, Christine (2000b), 'Re-examining Stardom: Questions of Text, Body and Performance', in Christine Gledhill and Linda Williams (eds.), *Reinventing Film Studies* (London, Arnold), pp.183-201.

Gledhill, Christine (2000), 'Rethinking Genre', in Christine Gledhill and Linda Williams (eds.), *Reinventing Film Studies* (London, Arnold), pp.221-244.

Gomery, Douglas and Allen, Richard (1985), *Film History: Theory and Practice* (New York, Mc Graw Hill).

Gorham, Maurice (1946), 'Television is Coming Back', in *The BBC Yearbook of 1946,* pp.18-21.

Gorham, Maurice (1949), *Television, Medium of the Future* (London, Percival Marshall).

Gorham, Maurice (1951), 'Television: A Medium in its own Right?', in Roger Manvell (ed.), *The Cinema of 1951* (London, Penguin Books), pp.131-147.

Gripsrud, Jostein (1998), 'Television, Broadcasting, Flow: Key Metaphors in TV Theory', in Christine Geraghty and David Lusted (eds.), *The Television Studies Book* (London, Arnold), pp.17-32.

Guback, Thomas (1985), 'Hollywood's International Market', in Tino Balio (ed.), *The American Film Industry* (Wisconsin, University of Wisconsin Press*)*, pp.463-486.

Gunning, Tom (1991), *D.W Griffith and the Origins of American Narrative Film: The Early Years at Biograph* (Illinois, University of Illinois Press).

Harbord, Jane and Wright, Jeff (1995), *40 Years of British Television* (London, Boxtree).

Hansen, Miriam (1991), 'The Return of Babylon: Rudolph Valentino and Female Spectatorship (1921-1926)', in *Babel and Babylon: Spectatorship in American Silent Film* (Massachusetts, University of Harvard Press), pp.243-269.

Harper, Sue and Porter, Vincent (1999), 'Cinema Audience Tastes in 1950s Britain', *Journal of Popular British Cinema* 2, pp.66-81.

Hill, John (1986), *Sex, Class, and Realism: British Cinema 1956-63* (London, BFI).

Hill, John (1996) 'British Television and Film: The Making of a Relationship', in John Hill and Martin McLoone (1996) (eds.), *Big Picture, Small Screen: The Relations Between Film and Television* (Luton, Luton University Press), pp.151-176.

Hill, John and Martin McLoone (1996) (eds.), *Big Picture, Small Screen: The Relations Between Film and Television* (Luton, Luton University Press).

Hilmes, Michelle (1989), 'The "Ban" that Never Was: Hollywood and the Broadcasting Industry in 1932', *Velvet Light Trap* 23, pp.39-47.

Hilmes, Michele (1990) *Hollywood and Broadcasting: From Radio and Cable* (Illinois, University of Illinois Press).

Holland, Patricia (1987), 'When a Woman Reads the News', in Helen Baehr and Gillian Dyer (eds.), *Boxed In: Women and Television* (London, Pandora Press), pp.55-63.

Holmes, Su (1998) 'Picturegoers and Tele-veiwers': A Study of *Picturegoer* 1950-1960', unpublished MA dissertation held at University of Southampton.

Holmes, Su (2001) 'As they Really Are and in Close-up': Film Stars on 1950s British Television', in *Screen*, 42:2, Summer 2001, pp.167-187.

Holmes, Su (2001) 'The Infant Medium with the Adult Manner': Television promotes the 'X' film (1952-62)', in *The Historical Journal of Film, Radio, and Television*, October, 2001, pp. 379-397.

Holmes, Su (2004) 'Looking at the Wider Picture on the Small Screen: Reconsidering Television and Widescreen Cinema in the 1950s', *Quarterly Review of Film and Video*, 21 (2) 2004, pp.131-147.

Holmes, Su (2004) 'Reality Goes Pop!: Reality TV, Popular Music and Narratives of Stardom in *Pop Idol*', *Television and New Media*, 5 (2), pp.147-172.

Holmes, Su and Deborah Jermyn (2004) (eds.), *Understanding Reality Television* (London: Routledge).

Hopkins, Harry (1963), *The New Look: A Social History of the Fifties and Forties* (London, Secker and Warburg).

Hutchings, Peter (2001) 'The histogram and the list: the director in British film criticism', *Journal of Popular British Cinema* (British Film Culture and Criticism), 4, pp.30-39.

Jacobs, Jason (2000), *The Intimate Screen: Early British Television Drama* (Oxford, Clarendon Press).

Jenkins, Henry (2000), 'Reception Theory and Audience Research: The Mystery of the Vampire's Kiss', in Christine Gledhill and Linda Williams (eds.) *Reinventing Film Studies*, (London, Arnold), pp.165-182.

Jewell, Richard B (1984), 'Hollywood and Radio: Competition and Partnership in the 1930s', *Historical Journal of Radio, Film, and Television* 4:2, pp.125-139.

Kerr, Paul (1996), 'Television Programmes about the Cinema: The Making of *Moving Pictures*', in John Hill and Martin Mc Loone (eds.), *Big Picture, Small Screen: The Relations Between Film and Television* (Luton, University of Luton Press), pp.133-41.

Kingsley Kent, Susan (1999), *Gender and Power in Britain: 1640-1990* (London, Routledge).

Klinger, Barbara (1991), 'Digressions at the Cinema: Reception and Mass Culture', in Patrick Bratlinger and James Naremore (eds.), *Modernity and Mass Culture* (Bloomington and Indianapolis, University of Indiana Press), pp.80-103.

Klinger, Barbara (1994), *Melodrama and Meaning: History, Culture, and the Films of Douglas Sirk* (Bloomington, Indiana University Press).

Klinger, Barbara (1998a), 'The Contemporary Cinephile: Film Collecting in the Post-Video Era'. Unpublished paper given at the conference *Hollywood and its Spectators: Reception of American Films 1895-1995*, University College, London.

Klinger, Barbara (1998b) 'The New Media Aristocrats: Home Theater and the Domestic Film Experience', *Velvet Light Trap* 42 (Fall), pp.1-19.

Kramer, Peter (1996) 'The Lure of the Big Picture: Film, Television and Hollywood', in John Hill and Martin McLoone (eds.), *Big Picture, Small Screen: The Relations Between Film and Television*, (Luton, Luton University Press), pp.9-46.

La Place, Maria (1987), 'Producing and consuming the Women's film: discursive struggle in *Now Voyager*' in Christine Gledhill (ed.), *Home is where the Heart Is: Studies in Melodrama and the Women's film* (London, BFI), pp.22-45.

Lafferty, William (1990) 'Feature Films on Prime-time Television', in Tino Balio (ed.), *Hollywood in the Age of Television* (Boston, Unwin-Hyman), pp.235-259.

Langer, John (1981), 'TV's Personality System', *Media Culture and Society* 4, pp.351-365.

Lavers, G.R and Rowntree, Seebohm (1951), *English Life and Leisure: A Social Study* (London, Longmans).

Leman, Joy (1987), 'Programmes for Women in 1950s British Television', in Gillian Dyer and Helen Baehr (eds.), *Boxed In: Women and Television* (London, Pandora Press) pp. 73-88.

Lewis, Peter (1978), *The Fifties* (London, Heinemann).

Lovell, Alan (2003) 'I Went in Search of Deborah Kerr, Jodie Foster and Julianne Moore but got Waylaid', in Thomas Austin and Martin Barker (eds.), *Contemporary Hollywood* Stardom, London: Arnold, pp. 259-270.

Macdonald, Paul (1998), 'Reconceptualising Stardom', in *Stars*, Richard Dyer (London, BFI), pp.177-180.

Macnab, Geoffrey (1993), *J. Arthur Rank and the British Film Industry* (London, Routledge).

Macnab, Geoffrey (2000), *Searching for Stars: Stardom and Screen Acting in British Cinema* (London,Continuum).

Maltby, Richard (1995), *Hollywood Cinema* (Oxford, Blackwell Publishers).

Mann, Denise (1992), 'The Spectacularisation of Everyday Life: Recycling Hollywood Stars and Fans in Early Television Variety Shows', in Lynn Spigel and Denise Mann (eds.), *Private Screenings: Television and the Female Consumer* (Minnesota, University of Minnesota Press) pp.41-69.

Manvell, Roger (1951), *A Seat at the Cinema* (London, Evans Brothers Ltd).

Marshall, P. David (1997), *Celebrity and Power: Fame in Contemporary Culture* (London, Minnesota, University of Minnesota Press).

Mc Donagh, Maitland (1996) *The Fifty Most Erotic Films of All Time: From Pandora's Box to Basic Instinct* (New York, Carol Publishing)

McArthur, Colin (2001) 'Two steps forward, one step back: Cultural struggle in the British Film Institute', *Journal of Popular British Cinema* (British Film Culture and Criticism), 4, pp.112-127.

McLoone, Martin (1996) 'Boxed in?': The Aesthetics of Film and Television', in John Hill and Martin McLoone (1996) (eds.), *Big Picture, Small Screen: The Relations Between Film and Television* (Luton, Luton University Press), pp.75-94.

Medhurst, Andy (1991) 'Every Wart and Pustule: Gilbert Harding and Television Stardom', in John Corner (ed.), *Popular Television in Britain: Studies in Cultural History* (London, BFI), pp.60-74.

Miller-Jones, Andrew (1948), 'Television and the Cinema', *Penguin Film Review* 6 pp.45-53.

Morley, David (1989) 'Changing Paradigms in Audience Studies', in Ellen Seiter et al (eds.) *Television Audiences and Cultural Power* (London: Routledge), pp.16-34.

Murdock, Graham (1992), 'Embedded Persuasions: The Rise and Fall of Integrated Advertising', in Dominic Strinati and Stephen Wagg (eds.), *Come on Down?: Popular Media Culture in Postwar Britain* (London, Routledge), pp.202-232.

Murphy, Robert (1989), *Realism and Tinsel: Cinema and Society in Britain 1939-49* (London, BFI).

Murphy, Robert (1992), *Sixties British Cinema* (London, BFI).

Park, James (1990), *British Cinema: The Lights that Failed* (London, Batsford).

Perry, George (1977) *The Great British Picture Show* (London, Hart-Davis).

Quinn, James (1968), *Film and Television as an Aspect of European Culture* (London, Leydon).

Neale, Steve (1990), 'Questions of Genre', *Screen*, 31: 1, pp.45-66.

Neale, Steve (2000), *Genre and Hollywood* (London, Routledge).

Neale, Steve (1998) 'Widescreen Composition in the Age of Television', in Steve Neale and Murray Smith (eds.), *Contemporary Hollywood Cinema* (London, Routledge), pp.130-141.

Negra, Diane (2002) 'Re-Made for Television: Hedy Lamarr's post-war star textuality', in Janet Thumim (ed.), *Small Screen, Big Ideas: Television in the 1950s* (London, I.B Tauris), pp.105-117.

Oswell, David (1999), 'And what might our children become? Future visions, governance and the child television audience in postwar Britain', *Screen,* 40:1, pp.66-86.

Parks, Lisa (2002) 'Cracking Open the Set: Television repair and tinkering with gender 1949-1955', in Janet Thumim (ed.), *Small Screen, Big Ideas: Television in the 1950s* (London, I.B Tauris), pp.223-243.

Paulu, Burton (1956), *British Broadcasting* (Minneapolis, University of Minnesota Press).

Paulu, Burton (1961), *British Broadcasting in Transition* (London, Macmillan).

Porter, Vincent (2000), 'Outsiders in England: The Films of ABPC, 1949-58', in Justine Ashby and Andrew Higson (eds.), *British Cinema: Past and Present* (London, Routledge), pp.152-166.

Pulleine, Tim (1982), 'Hollywood's Baby Brother?: British Films of the Fifties', *Films and Filming,* December, 1982, pp.24-29.

Quinlan, David (1983), *The Illustrated Guide to Film Directors* (London, Batsford).

Richards, Jeffrey (2000), 'Rethinking British Cinema', in Justine Ashby and Andrew Higson (eds.), *British Cinema, Past and Present* (Routledge, London), pp.21-35.

Robertson, James. C (1993) *The Hidden Cinema: British Film Censorship in Action, 1913-1972* (London, Routledge).

Robertson, Pamela (1996), *Guilty Pleasures: Feminist Camp from Mae West to Madonna* (London, I.B Tauris).

Robinson, Robert (1997), *Robert Robinson Memoirs: Skip All That* (London, Arrow Books).

Ryall, Tom (1996), *Alfred Hitchcock and the British Cinema* (London, Athlone).

Scannell, Paddy and Cardiff, David (1981), 'Radio in World War II' in *The Historical Development of Popular Culture in Britain 2* (Herts, Open University Press) pp.32-77.

Scannell, Paddy and Cardiff, David (1982), 'Serving the Public: Public Service Broadcasting before the War', in B. Waites et al (eds.), *Popular Culture: Past and Present* (London, Croom Helm) pp.29-40.

Scannell, Paddy (1990), 'Public Service Broadcasting: The History of a Concept', in Andrew Goodwin and Garry Whannel (eds.), *Understanding Television* (London, Routledge), pp.11-29.

Schatz, Thomas (1993) 'The New Hollywood', in Jim Collins, Hilary Radner and Ava Preacher Collins (eds.), *Film Theory Goes to the Movies* (London, Routledge), pp.8-36.

Sendall, Bernard (1982), *Independent Television in Britain, Volume One: Origin and Foundation* (London, Macmillan Press).

Shattuc, Jane (1997), *The Talking Cure: TV Talk Shows and Women* (London, Routledge).

Shaw, Colin (1999) *Taste, Decency and Media Ethics in the UK and USA* (Oxford, Clarendon).

Shearman, John (1946), 'Who are those Technicians?', in *Penguin Film Review* No. 6, pp.87-91.

Sinfield, Allan (1989), *Literature, Politics and Culture in Postwar Britain* (London, Blackwell).

Silvey, Robert (1974), *Who's Listening? The Story of BBC Audience Research* (London, Allen and Unwin).

Sobchack, Vivian (2000), 'What is Film History? or, the Riddle of the Sphinxes', in Christine Gledhill and Linda Williams (eds.), *Reinventing Film Studies* (London, Arnold), pp.300-315.

Spraos, John (1962), *The Decline of the Cinema: An Economist's Report* (London, George Allen and Unwin Ltd).

Spigel, Lynn (1992), 'Installing the Television Set: Popular Discourses on Television and Domestic Space, 1948-55', in Lynn Spigel and Denise Mann (eds.), *Private Screenings: Television and the Female Consumer* (Minnesota, University of Minnesota Press) pp.3–41.

Spigel, Lynn (1997), 'The Suburban Home Companion: Television and the Neighbourhood Ideal in Postwar America', in Charlotte Brunsdon, Julie D'Acci and Lynn Spigel (eds.), *Feminist Television Criticism* (Oxford, Oxford University Press), pp.211-233.

Spigel, Lynn (2001) *Welcome to the Dreamhouse: Popular Media and Postwar Suburbs* (Durham and London, Duke University Press).

Stacey, Jackie (1994), *Stargazing: Hollywood Cinema and Female Spectatorship* (London, Routledge).

Stafford, Roy (2001) '"What's Showing at the Gaumont?": Rethinking the Study of British Cinema in the 1950s', *Journal of Popular British Cinema* (4), pp.95-111.

Staiger, Janet (1991), 'Seeing Stars', in Christine Gledhill (ed.), *Stardom: Industry of Desire* (London, Routledge), pp.3-17.

Stempel-Mumford, Laura (1995), *Love and Ideology in the Afternoon: Soap Opera, Women and Television Genre* (Indiana, Indiana University Press).

Stokes, Jane (1999), *On Screen Rivals: Cinema and Television in the United States and Britain* (London, Macmillan).

Street, Sarah (2000), *British Cinema in Documents* (London, Routledge).

Street, Sarah (2001) 'The Idea of British Film Culture', *Journal of Popular British Cinema* (British Film Culture and Criticism), 4, pp.6-13. .

Studlar, Gaylyn (1996), '"Optic Intoxication": Rudolph Valentino and Dance Madness', in *This Mad Masquerade: Stardom and Masculinity in the Jazz Age* (New York, Rutgers University Press), pp.45-62.

Summerfield, Penny (1994), 'Women in Britain since 1945: Companionate Marriage and the Double Burden', in James Obelkevich and Peter Catterall (eds.), *Understanding Postwar British Society* (London, Routledge), pp.58-73.

Swann, Paul (1987), *The Hollywood Feature Film in Postwar Britain* (London, Croom Helm).

Swift, John (1950), *Adventure in Vision: The First 25 Years of Television* (London, John Leyman).

Thomas, Howard (1977), *With an Independent Air* (London, Weidenfeld and Nicolson).
Thumim, Janet (1991), 'The "Popular", Cash and Culture in the Postwar British Cinema Industry', *Screen*, 32: 3, pp.245-271.

Thumim, Janet (1998), 'Mrs Knight *must* be balanced: Methodological problems in researching early British television' in Cynthia Carter, Gill Branston and Stuart Allen (eds.), *Gender, News and Power* (London, Routledge), pp.91-104.

Thumim, Janet (2002a) 'Introduction: Small Screen, Big Ideas', in Thumim (ed.), *Small Screen, Big Ideas: Television in the 1950s* (London, I.B Tauris), pp.1-18.

Thumim, Janet (2002b) 'Women at Work: Popular Drama on British Television c1955-60', in Thumim (ed.), *Small Screen, Big Ideas: Television in the 1950s* (London, I.B Tauris), pp.207-222.

Todorov, Tzvetan (1977), *The Poetics of Prose* (Oxford, Basil Blackwell).

Uricchio, William, (1998) 'Television, Film and the Struggle for Media Identity', *Film History*, 10, pp.118-127.

Vahimagi, Tise (1996), *British Television: New Edition,* (Oxford, Oxford University Press).

Walker, Alexander (1983), *Joan Crawford: The Ultimate Star* (London, Weidenfeld and Nicolson).

Walvin, James (1978), *Leisure and Society: 1830-1950* (London, Longmans Press).

Wayne, Jane Ellen (1988), *Crawford's Men* (London, Robson Books).

Wegg-Prosser, Victoria (2002) 'This Week in 1956: The Introduction of current affairs on ITV', in Janet Thumim (ed.), Small Screen, Big Ideas: Television in the 1950s (London, i.B Tauris), pp.195-206.

Whannel, Garry (1992), 'The Price is Right but the Moments are Sticky: Television, Quiz and Game shows, and Popular Culture', in Dominic Strinati and Stephen Wagg (eds.), Come on Down?: Popular Media Culture in Postwar Britain (London, Routledge), pp.179-202.

White, Hayden (1975) Metahistory: The Historical Imagination in Nineteenth Century Europe (Baltimore, John Hopkins University Press).

White, Timothy (1990) 'Hollywood's Attempt at Appropriating Television: The Case of Paramount Pictures', in Tino Balio (ed.), Hollywood in the Age of Television (Boston, Unwin-Hyman), pp.145-165.

Williams, Raymond (1983) Keywords (London, Fontana).

Williams, Raymond (1974), Television: Technology and Cultural Form (London, Collins).

Wyndham Goldie, Grace (1977), Facing the Nation: Television and Politics 1936-76 (London, Bodley Head).